TINS
-DECORATIVE PRINTED-

THE GOLDEN AGE OF PRINTED TIN PACKAGING

DAVID GRIFFITH
RESEARCH: CHARLOTTE PARRY-CROOKE

1979
CHELSEA HOUSE PUBLISHERS
NEW YORK, LONDON

ACKNOWLEDGEMENTS

To my friend Roy Morgan, who encouraged me to be the first to compile a book on the very best of British printed tins.

Assembling the material for this book would have proved impossible had it not been for the invaluable help and co-operation of the following, including those who most generously loaned their possessions for our use:

Ian Allen; Christopher Ashton and Anjula Daniel; John Badcock; John Battershell; Ken Carter; James C Collins; J & J Colman: Reg Butcher; Stuart and Linda Cropper; John and Kathy Emmett; Elizabeth Farrow; Malcolm Gliksten; Honor Godfrey; Christopher Gordon; Kenneth Griffith; Richard E Hale; Jack Haythornthwaite; Huntley & Palmers: Michael Paxton and Kate Hogg; Imperial Group: Brian Freeman and his department (Patents and Trademarks); Diana Vasavour James; Eric King; Ian Logan; Metal Box, Carlisle: Herbert Armstrong; Metal Box, Mansfield: John Scott and Stephen Radford; Metal Box, Reading: Edward Garling and Peter Bartlett; Brian Morgan; Godfrey Omer-Parsons; Robert Opie; Aldo Osti; John Player & Sons: Alan Dobson; Katy and Johanna Pottle; Reading Museum and Art Gallery: Sue Reid; Reading University: Mr. Edwards and Professor Twyman; Bert Rumely; Robert Scully; Ken Sequin; Alan Starkey; Ian and Rita Smythe; Audrey Tester; United Biscuits: Doris Flintham; Diana de Upagh; Judy Walker; Brian Webb; W.D. & H.O. Wills: David Redway and Hubert Rudman; Patsy Wood.

All the photographs were taken by Norman Hollands except those taken or provided by the following: J & J Colman/Reg Butcher 19 (top and bottom right), 20 (bottom), 62 (bottom), 63 (bottom), 64 (bottom), 65 (bottom), 80 (bottom), 82 (top); Adabelle Davis 90 (top); Richard Hale 44 (bottom); Huntley & Palmers 28 (top); Metal Box/Bob Enever 13, 37 (Bottom), 41 (top far left and bottom), 42 (bottom right), 49 (bottom far left), 51 (top), 52 (bottom left), 68 (bottom), 70 (top), 74 (bottom); Metal Box, Mansfield 14 (bottom), 17, 20 (top); Metal Box, Reading 14 (top and centre), 19 (left); John Player & Sons 72 (top right); Reading University Library 84 (bottom left); Ken Sequin 26 (bottom left), 54 (bottom); Peter Simpkin 90 (bottom), 91 (top and bottom), back jacket flap

I would also particularly like to thank the following: Norman Hollands for his painstaking attention to quality and detail in the large proportion of photographs taken by him; Josephine Wright for typing the original manuscript, and for providing encouragement whenever required; Reg Boorer for his assistance and contribution to tasks over and above his official duties as designer, and in particular his effective arrangements of the contents of the photographs; and Charlotte Parry-Crooke, whose tireless efforts in every aspect of the book have contributed immeasurably to its authority.

Editor: Charlotte Parry-Crooke
Designer: Reg Boorer
Photographer: Norman Hollands

Copyright © 1979 David Griffith, Reg Boorer and Charlotte Parry-Crooke
First published in 1979
All rights reserved. No part of this
publication may be reproduced, stored in
any retrieval system, or transmitted in any
form or by any means, electronic, mechanical,
photocopying or otherwise,
without the prior permission in writing of
the Publishers.

ISBN: 0-87754-095-0 (cloth)
ISBN: 0-87754-099-3 (paper)
LC: 79-52192

CHELSEA HOUSE PUBLISHERS
Harold Steinberg, Publisher & Chairman
Andrew E. Norman, President
Susan Lusk, Vice President
A Division of Chelsea House Educational Communications, Inc.
70 West 40th Street, New York 10018

Contents

Introduction. Page 7

Printing on Tin. Page 11

Huntley & Palmers. Page 21
Pioneers of imaginative collectable
decorative tins for biscuits

Biscuit Tins. Page 35
Fine packaging from over ten
of Huntley & Palmers' competitors

Confectionery. Page 47
Distinctive tin containers for sweets, toffees,
cachous, chocolates and cough lozenges

General Provisions. Page 55
Nostalgic reminders of grocery tins which
filled the shelves of the corner shop

Cigarettes and Tobaccos. Page 67
Eye-catching advertising of the competitive
tobacco industry

Commemorative Tins. Page 77
Glorious events of British history
immortalized on tin

Collections. Page 83
Specialist collecting and collectors in
Britain and America

Tin Identification Index. Page 94

General Index. Page 96

– INTRODUCTION –

I HAD ONE OF THOSE BUT I THREW IT AWAY — that's the name of an antiques shop spotted while driving through a small English country town and, having been a collector of tins for over twelve years, the familiarity of the words on that fascia board rang a most promising bell. It is a cry that is frequently heard by anyone involved with old advertising art, and is itself one of the reasons for the increasing popularity of collecting printed tins. Who, after all, is likely to find a Rembrandt in the attic? But there might well be a Sharp's *Toffee Bucket* or a Huntley & Palmers miniature originally picked up on a visit to the Festival of Britain.

The growing interest in printed tins and advertising art in general took root as recently as the last decade. As a reaction to the post-War austerity of the 1940s and 50s, the 'swinging' 60s heralded a definite reversal in taste. Hitherto unfashionable Victoriana began to fill antiques shops and stalls at prices that people could afford, and Victorian buildings such as railway stations, pubs and old shops started to be appreciated rather than dismissed or demolished unquestioningly. Those people who had been indoctrinated with the theory that Victorian taste equalled bad taste were now attracted by the exuberance and decorativeness of the very things that they would have previously sent to the junkyard, suddenly realizing the need to preserve pub mirrors, tiles, signs, printed tins and many other nostalgic reminders of an age of fine craftsmanship.

Accompanying these changes in taste came the Sixties craze for military nostalgia. Of the patriotic tins produced, some were not newly commissioned but copies of designs that had originally been produced some fifty or sixty years previously. Inevitably, as these reproductions became more popular, interest in the genuine articles started to increase, the search for old tins given added momentum by the general nostalgia boom.

When in 1971 an exhibition of decorative biscuit tins was held at the Victoria and Albert Museum, London, the validity of the printed tin became established as an art form in its own right. Since then further exhibitions have been mounted, notably the large exhibition of all types of packaging drawn exclusively from the fine collection of Robert Opie (see page 91), also held at the Victoria and Albert, and most of the auction houses now regularly hold 'collectors' sales of all forms of commercial art. Reproductions of many classic containers are widely available, as are newly designed examples, all contributing to the growing popularity of printed tins.

Lord Kitchener shown here was featured on the most famous patriotic reproduction tin of the Sixties, and even gave his name to a shop in London's Carnaby Street, 'I was Lord Kitchener's Valet'.

In America, too, collecting old advertising art is popular and collectors are keen to acquire British tins, the evocative advertising images of Victorian and Edwardian Britain and the quality printing and construction of the containers contributing to their appeal. As in Britain, reproduction tins are available and nostalgic 'country store' antiques shops abound, as well as 'old advertising shows' which are held in different states, drawing thousands of buyers.

Given the increasing interest in tin collecting, it is surprising that no publication has yet been produced on British printed tins, for, whether we are attracted by the elegant graphics and decorative qualities, the impressive skill and craftsmanship involved in their construction, their documentation of social history, or by the fact that they were designed to appeal to the man in the street and as such are fascinating reflections of their times, printed tins can appeal to us all.

Different people of course find different aspects of tins attractive but one characteristic of their appeal which always seems to feature prominently is the high standard of craftsmanship achieved. Although there were thousands of utilitarian tins produced which are not now collected, these standards of craftsmanship are evident in a large proportion of the tins produced throughout the timespan — roughly 1860-1940 — of this book. Although tins were mass-produced in large numbers, each stage of their manufacture, from the design to the printing and assembly, was carried out by expert craftsmen to standards that cannot now be achieved.

Whilst no-one can help but marvel at the staggeringly high standards achieved, personal preferences draw people in different directions when buying and collecting. Some collectors are drawn to novelty tins in the shapes of real objects, in which field the designers' imagination seemed inexhaustible, while others favour the 'shaped' tins with their nostalgic illustrations of British history and the Empire, containers admired for the brilliance and depth of their colour and attention to detail. For others bold graphics are the attraction, advertising tins often bearing a bewildering variety of ornate hand-drawn lettering on the same tin, not to mention boastful claims made for the products contained inside. Many containers can also aid the study of individual companies or commodities, as well as giving insights into the fashions and fads of the years in which they were produced, and tins of all kinds make attractive displays.

At the time of writing, little written information is available on the processes involved in the production of printed tins, or on their issue by the different companies (with the exception of Huntley & Palmers), but hopefully with time more facts will come to light. This book therefore is intended as only an introduction to the golden age of British decorative printed tins.

Examples of different types of tin that appeal to different collectors. From top to bottom: Jacob's inventive novelty *Wishing Well*; Keen Robinson's decorative mustard tin *National Game*; McVitie & Price's patriotic commemorative *Victoria Cross Scenes*; Clarnico's miniature *Teddy Bear*, and two good examples of advertising tins with fine graphics and illustrations.

Where to Look for Tins

The search for printed tins is bound to lead those interested into some unlikely places, since as yet there are few specialist dealers. More and more frequently, tins are to be found on the shelves of the general antiques shop, particularly those that stock other related advertising items. For some years 'collectors' sales' at leading auction houses have included a sprinkling of tins, and in October 1978 a major collection

INTRODUCTION

of tins was auctioned by Christie's, the first sale of its kind.

Specialist rallies and ephemera fairs can also prove productive hunting grounds, but flea markets and local antiques fairs are still the most likely places to yield plentiful supplies of tins. Second-hand and junk shops, and dealers who specialize in house clearances are also good sources. There is always the chance of finding a box of old tins from a recently-cleared pantry or boxroom. Attics often yield incredible finds — novelty toy tins stored there since childhood — and gunrooms, garden sheds and garages are quite likely to produce tins now used for storing nails and so on.

Though it is difficult to make any rigid rules, the North of England and other traditional industrial regions tend to be particularly rich in tobacco tins, as well as confectionery and general provisions containers. The more elaborate, and comparatively more expensive, biscuit and mustard tins are more plentiful in middle-class areas.

ABROAD, and particularly in the USA, British tins are to be found in increasing numbers in antiques shops and markets, and reproduction country stores now offer British as well as American tins. Many of the British items on sale are bought up in Britain by foreign dealers for re-sale, but tins discovered in a local garage sale or on the shelves of the real old-fashioned country store may well be examples of the quite significant proportion of tins which were originally manufactured for export. Agents and representatives were employed by companies in different countries to handle their interests, and in Huntley & Palmers' case, catalogues were printed in six languages.

Condition and Valuation

Since most tins were designed for use rather than for posterity, it is very rare to find examples in truly pristine condition. Ironically, when this does happen, the pristine tins tend to look rather garish. Although there is an inevitable amount of compromise in acquiring a tin, a point is reached when damaged or rusted surfaces render even a rare and otherwise desirable container virtually worthless. This stage is generally agreed to have been reached when the state of the tin really detracts from its original appeal.

A tin's condition is one of the major factors in establishing its value, but valuation is an aspect of tin collecting about which it is difficult to give specific details. When any type of collectable object rapidly gains popularity prices tend to fluctuate irrationally. This has certainly been the case with tins. It would therefore be both impractical and misleading to compile any form of price guide, particularly since there is a considerable gulf between prices paid in Britain and abroad.

This gulf is most evident in North America where demand exceeds supply and prices are therefore forced up. On the other hand, the notion that the antiques markets of Britain are laden with high-quality tins at bargain prices has in reality proved a bitter disappointment to many a foreign collector. Given these two sides to the coin, movements are now beginning to be made in the direction of establishing an agreed international price level.

An interesting difference between printed tins and other antiques fields is that rarity plays a less significant role than usual in the desirability, and therefore the value, of tins. Even when a hitherto undocu-

Places in Britain to Look at Old Advertising Material

Bath, Avon
The American Museum in Britain at Claverton Manor
(reproduction country store)

Brighton, East Sussex
Royal Pavilion Art Galley and Museum – Seaside Gallery

London
Museum of London, London Wall
(reproduction shops)

Manchester
Whitworth Art Gallery,
University of Manchester

Norwich, Norfolk
The Mustard Shop
(J. & J. Colman Ltd)

Oxford, Oxfordshire
Bodleian Library
John Johnson Collection
(The Sanctuary of Printing)
(by special application only)

Reading, Berkshire
Reading Museum and Art Gallery
(Huntley & Palmers room to open shortly)

Stanley, County Durham
Beamish North of England Open Air Museum

York, North Yorkshire
Castle Museum (reproduction shops)

A Selection of the Many Places in the U.S.A. to Look at Old Advertising Material

Buena Park, California
Knotts Berry Farm
general store

Cooperstown, New York State
Cooperstown Museum

Shelburne, Vermont
Shelburne Museum

Sturbridge, Massachusetts
Sturbridge Village Country Store

Washington, D.C.
Smithsonian Institute

Associations to Join

Ephemera Society
12 Fitzroy Square
London WC1
(publishes *The Ephemerist*)

History of Advertising Trust
53 Goodge Street
London W1
(publishes *Journal of Advertising History*)

Tin Container Collectors Association
P.O. Box 4555
Denver, Colorado 80204, U.S.A.
(publishes *Tin Type*)

Further Reading

Adburgham, Alison
Shops and Shopping 1800-1914
Allen & Unwin, London, 1964

Alford, B. W. E.
W. D. & H. O. Wills and the Development of the U.K. Tobacco Industry 1786-1965
Methuen, London, 1973

Clark, Hyla M.
The Tin Can Book
New American Library, New York, 1977

Corley, T. A. B.
Quaker Enterprise in Biscuits: Huntley & Palmers of Reading 1822-1972
Hutchinson, London, 1972

Davis, Alec
Package and Print: The Development of Container and Label Design
Faber & Faber, London, 1968

Dover Books Pictorial Archive Series on Graphics, Packaging and Crafts
Dover Publications, New York

Grossholz, Roselyn
Country Store Collectibles
Wallace-Homestead Book Company, Des Moines, Iowa, U.S.A., 1972

Jolley, David and Mullen, Chris
Yellow, White and Blue: The Advertising Art of J. & J. Colman Ltd.
J. & J. Colman, Norwich, 1977

Jones, Owen
Grammar of Ornament (facsimile edition)
Van Nostrand Reinhold, London and New York, 1972

Kirk, Albert
Metal Box Decoration
Printing Craft Lecture given 21st January, 1938 at Stationers Hall, London
Printing Trade Lectures, series 16, no. 4

Oddy, Derek J. and Miller, Derek S.
The Making of the Modern British Diet
Croom Helm, London, 1976

Petit, Ernest L.
Collectible Tin Containers
Forward's Color Productions, Manchester, Vt., U.S.A., 1970

Rawlins, Chris
English Biscuit Advertising and Tins
Collectors Weekly, Kermit, Texas, U.S.A., 1974

Reader, W. J.
Metal Box: A History
Heinemann, London, 1976

Turner, E. S.
The Shocking History of Advertising!
Michael Joseph, London, 1952;
Penguin Books, Harmondsworth, Middx, 1968

Twyman, Michael
Printing 1770-1970
Eyre & Spottiswoode, London, 1970

mented tin is found, if it is unattractive and of uninspiring subject matter, it can be considered worthless. Conversely, tins such as Huntley & Palmers' book sets which abound on the shelves of antiques shops, are regularly sold for high prices.

Care of Tins

Like most antiques, tins need to be properly cared for if they are to remain in good condition, and it is sensible to display them out of direct sunlight as the colours can, and do, fade. More important, however, is to make sure that they are kept out of steamy atmospheres such as kitchens and bathrooms; placed there they will court disaster from rust. Once a tin has rusted externally there is little that can be done to remedy its condition, since the oxidization of the metal will have destroyed the printed image. Internal rusting is not as serious but it should be treated immediately, either with soft wire wool or with a wire brush, as once the corrosion has taken root it can eat through to the printed surfaces.

Dents are another often under-estimated cause for concern. The pressure of the impact that resulted in the dent will have stretched the metal, and the tin can never be returned to its original shape. However, except in the case of novelty tins, which often have separate tin linings, some improvement can usually be made. The method most favoured by collectors is to take a rounded wooden spoon or spatula and with it apply firm strokes with even pressure on the underside of the dent until it is flattened out.

Grease and dirt can usually be removed from a tin's surface without causing harm to the printed image. Each collector has his own favoured cleaning solution, and non-abrasive liquid metal polishes, car colour restorers or mild soap and water are all popular methods. However great care must be taken, as the different printing processes have variable resistances. Many a fine transfer-printed tin has had its design wiped away at a stroke by an over-enthusiastic collector, so it is imperative to test a small corner of the tin first. To give both lasting protection and an attractive finish to a newly restored tin, a coat of wax polish can be applied, but never use varnish. It will discolour the surface, and cannot be removed without destroying the printed image. Finally, when transporting tins, it is a good idea to wrap them in several sheets of newspaper to avoid scratching.

– PRINTING ON TIN –

A Brief History of Decorative Tin Printing Developments

T HE TIN PRINTING INDUSTRY IN BRITAIN has always had a reputation of secrecy, and it is without doubt a justified one. Not only is the early history of inventions and developments for printing on tin largely undocumented except in patent specifications or trade journals, which refer to intrigues, mysterious figures and secret presses, but the actual day-to-day skills and craftsmanship, and intimate knowledge of the many staggeringly complicated processes used in the production of printed tins during Victorian and Edwardian times also remain for the most part unrecorded.

Nonetheless, due particularly to the detective work of two or three individuals (see Further Reading), the remarkable story of the chronological developments in tin printing from the early nineteenth century can to some degree be pieced together. Not content with the lack of factual detail, it is a tale further complicated by the use of confusing terminology, in some instances two completely different processes being described by exactly the same term.

While printers were from the early nineteenth century continuously (and usually secretly) experimenting with new ways of printing successfully on tin — a notoriously unreceptive and difficult material on which to print — or adapting and improving already accepted methods, the most important developments break down into three basic chronological areas and concern several main companies. 'Direct' printing was in use from the mid-1850s, involving the Tin Plate Decorating Co in particular; this gradually gave way during the late 1860s and 70s to 'transfer' printing which was used by many companies, but which was eventually superseded by 'offset lithographic' printing (also confusingly called 'direct' printing), invented specifically for printing on tin but not immediately available for general use. The names most often associated with this last development are Barclay & Fry and Huntley, Boorne & Stevens. The two other large printing firms that play an important part in this story are Hudson Scott of Carlisle and Barringer, Wallis & Manners of Mansfield.

The use of all three methods, due to strict patent restrictions, the expense of installing new machinery and the conservatism of many firms, overlapped each other and were indeed quite often combined on tins. All however relied on one major printing principle. This was lithography, and it is impossible to discuss the three tin printing techniques without continuous reference to the lithographic and chromolithographic printing processes.

Two early Huntley & Palmers biscuit tins, both transfer printed by Ben George. The one on the right, designed by Owen Jones and produced in 1868, is the earliest known printed biscuit tin.

A lithograph by Quaglio of Alois Senefelder, the inventor of lithography.

Invented in Munich in 1798 by Alois Senefelder, lithography was the first 'planographic' printing process to be developed. It was so-called because, unlike other methods — etching, wood engraving and so on — which relied on printing the image from a relief or intaglio design, both printing and non-printing areas in a lithograph were on the same 'plane' or level. This was achieved by the application of the principle that grease and water do not mix, yet both are accepted by a porous substance such as limestone. A design drawn on limestone in a greasy substance will repel water, which will however be absorbed by all other areas of the porous stone. Consequently an application of a greasy printing ink will adhere to the greasy design but not to the damp areas, and the design can therefore then be transferred to the receiving vehicle by passing it through a press.

For many years black printed lithographs and other types of print were coloured by hand and it was not until towards the mid-nineteenth century that processes for actually printing in full colour were developed. Of those invented 'chromolithography' quickly became the most popular and accessible. Patented in Paris in 1837 by Godefroye Engelmann, 'chromolithographie', which simply means 'printing in colours by lithography', involved the coloured image to be depicted being analysed into the various separate colours of which it was composed (hence separations). The areas of these separate colours were then drawn by hand onto as many different stones as there were colours, sometimes up to 12 or 14, in addition to the black 'key' stone. From them the lithograph would be accumulatively built up, colour by colour. This was the method of colour printing employed by the tin printer, and the use of its detailed and effective processes occasioned the production of some of the most beautiful examples of transfer and offset printed tins.

Until the gradual, but somewhat reluctant, acceptance by the tin printing industry in the 1920s and 30s of mechanical (as opposed to hand) separations into only four basic colours by photolithography, chromolithography was consistently used for printing on tins in full colour from the 1870s. It should however be remembered, to avoid any confusion, that this was the method by which coloured images were separated and then prepared for printing, rather than the process by which they were actually transferred to the tin plate.

The first stage of the tin printing story — 'direct' printing — is, however, still surrounded by a certain amount of confusion and by all accounts it involved some strange adaptations of regular printing processes. While printers experimented in the 1840s and 50s with different ways of printing on tin, they came up time and again against two basic problems — there was on the one hand the total inability of sheets of non-porous tin plate to absorb printing inks, and on the other, the lack of sympathetic contact between the hard and unyielding tin plate and equally solid printing vehicles such as stones or blocks. These two difficulties account for the comparatively late development of successful printing on tin, and it was only with the subsequent introduction of an intermediate stage which removed the need to place the tin plate directly in contact with the stone that these obstacles were eventually overcome.

The most successful early attempts at direct printing were made at the Tin Plate Decorating Co. in Wales which shared premises with Leach, Flower & Co., tin plate manufacturers. They tried various processes, including complicated combinations of etching and lithography in which images on lithographic stones were etched up for direct contact with tin plate and printed in black only. Success was not surprisingly somewhat limited; results achieved were uneven, and expensive lithographic stones subjected to great pressure were often cracked or broken in printing.

Despite their efforts the Tin Plate Decorating Co., although continuing to use 'direct' printing, adopted alongside it a much more satisfactory and effective method, favoured by other printers, of printing images and designs onto tin, this time in full chromolithographic colour: transfer printing. This was a process similar to that used on Battersea enamels and it involved lithographically printing in reverse on to thin paper designs or illustrations, which were then 'transferred' under pressure to the tin plate in a lithographic transfer press. The paper was then soaked off in a water bath to leave complete and perfect illustrations on the tin plate.

Again the date of when transfer printing onto tin was first used in Britain is uncertain, but there was a good deal of communication with French printers on the matter from the 1860s. France led the field in decorative printing, and British printers visited the Continent to study advanced printing methods and to order beautifully printed colour 'decalomanie' transfers from them for use on British-made tins.

In Britain, however, the name most often associated with successful developments in transfer printing is that of Benjamin George, a printer who worked in Hatton Garden, London, from 1856 to 1893. A prolific patentee of printing processes, he was the pioneer in Britain of printing by the transfer method — the first really successful technique of printing on to tin in full colour. He invented several ways of improving quality yet saving time, including coming up in 1870 with a new way to prepare tin plate for printing. He proposed adding final extra coats of flat colour to the transfer, printed last to come off first on the tin, thus avoiding the need to coat, dry and rub up the actual sheet by hand in advance, each stage of which often had to be repeated several times. The patent also mentioned the use of clear lacquers on tin to achieve metallic effects, and the use of embossing to give 'extra richness to the design'. His later patent involved increasing the speed of transfer printing by sandwiching transfers and tin sheets between cardboard in piles, so that as many as 40 sheets, subjected to roller pressure, could apparently be printed at once.

George's main claim to fame however is that his name appears as patentee on a Huntley & Palmers biscuit tin of 1868 which has been described as 'probably the earliest surviving example of tin printing to which a definite date can be ascribed' in Britain. This was the Huntley & Palmers *Casket* tin, designed by the celebrated Victorian designer, Owen Jones, in commemoration of the granting of royal patronage to the company. For this and other transfer-printed tins by him, George supplied the tin box makers, Huntley, Boorne & Stevens with flat printed sheets for which he was paid in the case of *Casket* (a 12-13 colour job) £3.10s for every 100, the boxes themselves being made up in Reading.

This very early mustard tin was transfer printed in black, with gold lacquer and colour printed paper insets. It was produced by the Tin Plate Decorating Co. for Barringer & Brown, who subsequently sold out their mustard-making business to Colman's, to concentrate, as Barringer, Wallis & Manners, on tin printing and box making.

Robert Barclay, the inventor of offset lithography, photographed in about 1850.

John Doyle Fry, Barclay's partner and co-patentee in the first offset lithography patent of 1875.

Right: Barringer, Wallis & Manners' superb display lorry, on which can be seen several of the tins featured elsewhere in this book, including the Victory-V clock set.

The transfer-printing process was used by most other tin printers until after the patent expiry in 1889, which allowed them to take advantage of an invention which ultimately revolutionized both paper and tin printing industries, but one which was specifically invented for printing on tin: offset lithography.

Developed by the London printing company of Barclay & Fry (who were connected with Barclays, the bankers and Fry's, the confectionery firm), their method overcame the problem of unsympathetic direct contact between stone and tin plate, and it also avoided the lengthy intermediate paper transfer stage, these advantages causing it to be also known most confusingly as 'direct' printing. The entire printing process could now take place on the press without stone and tin meeting, since the image on the litho stone in the press could be 'picked up' by a non-absorbent glazed cardboard impression cylinder (later changed to a rubber blanket) and then immediately offset onto the tin plate being passed through the machine.

The invention was patented in 1875; a second patent taken out in July named Barclay only, but the earlier March one named both Barclay and Fry, as well as referring to a 'communication from abroad from Henry Baber of Paris'. Known for years as the mysterious Frenchman, Baber was in fact an English printer who had just returned from Paris, parting with some of his newly acquired information to Barclay & Fry for the meagre sum of £5. For a variety of reasons, including lack of funds (despite banking connections), to finance the re-organization of machinery, the patent rights were sold in 1877 to the match manufacturers Bryant & May. They in turn

licensed the tin box makers, Huntley, Boorne & Stevens, who had already turned down the rights themselves, to operate the process, financing for them a printing works complete with offset machinery. Huntley, Boorne & Stevens thereby became the first company in the country to combine tin box making with tin printing and, with Huntley & Palmers as their main customers, they were set for success.

Other companies, including Barclay & Fry themselves, rushed to adopt the process when the patents lapsed in 1889, and rumours abound that some firms had, prior to that date, already been experimenting with the method, in particular Hudson Scott. The remarkable Mr Baber had moved to Carlisle as their manager in 1877, and while they admitted, in a patent infringement episode with Bryant & May, to experimenting with new processes from the Continent for 'direct' tin printing (which must surely have been offset tin printing given Baber's knowledge), they were 'officially' still printing by transfer until after 1889. Barringer, Wallis & Manners also took up the process as soon as they could — certainly before 1893.

Offset printing machinery quickly became more sophisticated. In the original 'flat bed' presses, stones lay face up in the press and tin plate sheets were fed in between the rollers face down, so that the printed image could not be seen until it came out of the press. Just after 1900 metal litho plates began to be widely used instead of stones, and this occasioned the invention by Geo. Mann & Co., printing machinery manufacturers, of a new type of press, the Mann Standard rotary, into which sheets were fed vertically, the printed side visible at all times. Its output was double that of the flat bed and it was such a success that it apparently caused a 'furore' at the 1904 Printing Machinery Exhibition at the Agricultural Hall, Islington. It was for many years after the most popular tin printing press.

Luckily for the collector of printed tins, and the tin printing enthusiast — for the quality of chromolithographically printed tins is unquestionable — the tin printing industry was slow to adapt to the use of photolithography for the reproduction of designs and illustrations. It was only in the late 1920s that either Hudson Scott or Barringer, Wallis & Manners started to use this method and it was for many years employed side by side with chromolithography.

The Production of a Decorative Printed Tin

Any attempt to describe the many varied and complicated stages through which a decorative tin must pass from concept to its final appearance on the shelf of the corner store is a daunting undertaking. While specific information on the actual printing of tin plate sheets is available (thanks to Albert Kirk's fascinating 1938 lecture on Metal Box Decoration), other important stages that a tin would have passed through in its production remain virtually undocumented, and much of the story can consequently only be pieced together in a conjectural way.

For an industry which catered for a mass market, the tin printing trade impressively still relied in the late nineteenth and early twentieth centuries on the capabilities of individual craftsmen at many stages of the production line. While some of the jobs were of course mechanical, others depended totally on expert and often intuitive

Chromolithographed pages from Owen Jones's *Grammar of Ornament* (1856), a standard reference book for designers and artists for many years.

knowledge. Apprentices, taken on at an early age, spent years in one department acquiring these skills under the practical guidance of the head printer, varnisher or boxmaker, whose own knowledge was handed down by word of mouth. Instruction manuals were of course produced but, few and far between, they are nowadays almost unintelligible to the layman, referring blithely to tools, processes and terminology now obsolete. On the other hand, visual reference books and guides to colour, lettering, and decoration abound, and these were certainly in constant use in tin printing firms' studios where the life of a tin began.

At Hudson Scott the 'Artist Room', as it was known, encompassed both the creation of designs and illustrations and their 'reproduction' onto litho stones ready for printing, and while chromolithography was still in use, the foreman of the department took each tin through from concept to proofing stage. It was usual for printed tin designs to be prepared in the creative side of the studio, sometimes the customers themselves proposing ideas, as Fryer's (Victory-V) certainly did. Usually wooden working models of the tins, decorated with rough paper illustrations, were constructed and put forward for approval. As was to be expected, most ideas relied heavily on current artistic and printing fashions for their inspiration. At Hudson Scott apprentices were taken to London exhibitions and 'to look at shop windows for ideas for designs', visits were made to the Continent by senior designers, and over the years a considerable reference library was built up. This would no doubt have included many volumes of designers' aids — ornamental borders, strapwork, scrolls, rules and other decorative items — themselves often beautifully printed in chromolithography — such as Owen Jones's justifiably well-known *Grammar of Ornament* (1856 and subsequent editions), as well as reference for the ever increasing range of lettering styles and display typefaces, influenced both by late nineteenth century 'artistic' and 'antique' printing fashions, and then Art Nouveau, and also by contemporary sign writing and shop front lettering.

Different designers and artists would more than likely have specialized in one of the different types of tin produced, decorative tins such as biscuits and mustards, which featured fine illustrative scenes to be undertaken by the more 'artistically' inclined. Intended specifically for attractive after-use, these tins were particularly ornate and lavish in their use of decorative detail. Floriated borders and repeat geometric patterns enclosed 'artistic' panels beautifully drawn in many colours, depicting favourite themes.

After the turn of the century novelty tins would have claimed the attention of the most imaginative and inventive designers. In place of flat graphics and illustrations came every sort of three-dimensional object, from laundry baskets to garden rollers, used as inspiration for novelty containers. In this field the skill lay not so much in the preparation of the images in the artists' department, as in the adaptability of the tool and box makers who had actually to invent the individual dies by which the newly designed tins could be stamped out and constructed.

Scope for two-dimensional commercial art came in the advertising tins produced to package ordinary provisions. Although the most basic utilitarian shapes were used, designers had a field day with bold, blatant and often unconventional graphics combined with beautiful imagery, decorative scrolls and fancy borders. Lettering in particular was used in a multitude of forms and sizes, and since with lithography there were no mechanical type restrictions, everything being drawn up by hand, artists went to town adding flourishes to letters and taking great liberties with standard alphabets. Product names could be bent, squashed or stretched according to whim, with initial capital letters usually used much larger than the rest of the word. Especially popular were rustic lettering styles, including those in the forms of twigs and branches as well as those drawn as rope, and shadow and three-dimensional letters.

Artists working on litho stones in the studio at Barringer, Wallis & Manners. Examples of their work, including a Player's 'lifebelt', can be seen on the walls.

Once a final design for any of these different types of tin was ready, it was prepared for printing. First of all it had to be 'separated' into as many colours as were required and then transferred onto a similar number of litho stones from which the tin printing was done. This process appears, at least at Hudson Scott, to have been the responsibility of the 'reproducers' in the Artists Room.

When the number of colours in which the design was to be printed had been decided upon, quite often as many as twelve or more, the design had to be analysed into these separate individual colours. This was one of the most important stages through which a tin design had to pass, and it was an exceptionally tricky and skilled business. Technical manuals, such as Field's *Grammar of Colouring*, stipulated rules for colour harmonies and contrasts, illustrating all the combinations of the vast range of colours and tints to choose from, and reproducers had to be able to forsee all the possible combined printed effects of the colours chosen.

Whether the individual colour areas were then drawn directly onto the stone, as in the early days, or more conveniently from about 1870, onto transfer paper (this is a completely different process from 'transfer printing') and thence put onto the stone, this colour separation stage had always to be completed. If the image was to be drawn directly onto the stone, the outline was 'traced down' in reverse. From this 'key' stone, on which the outline image and register marks had been drawn, prints were taken and offset onto as many stones as there were colours. Similarly, if the transfer method was

DECORATIVE PRINTED TINS

A detail of an image separated by modern photolithography into only four colours, as compared with chromolithographic separation into as many as 12 or 15 different colours.

Right: A fine example of chromolithography: an enlarged detail of the *Who Killed Cock Robin?* nursery rhyme tin from Farrows, showing the separate 'dots' of colour. The whole tin is featured on page 65.

being used, which was much easier to work with since the separations could be worked up the right way round rather than in reverse as on the stone, the image outline and register marks were printed onto as many pieces of transfer paper as there were colours.

The stones or transfers for the individual colours were then worked up by techniques adapted for the two different methods. On stones, grained effects for gradations of colour were originally achieved by using chalk on the grained stone, while an embossed grain was applied to transfer paper, and then chalked to get the same result.

There was one particularly important technique associated with chromolithographic work — stippling. A process already used much earlier, stippling was revived when graining was no longer possible due to the highly-polished stones that had to be used in the new fast presses. It involved the use of dots to interpret the colour values and it was similar to the techniques of neo-impressionism and photolithography, both of which it anticipated. In this method each colour, lighter ones first, was stippled to various depths with a fine steel pen or brush in minute dots. It was a very laborious process, especially if smooth regular effects were to be achieved. In order to alleviate this complex and time-consuming labour, semi-mechanical aids such as 'Day's Shading Medium' were introduced, similar in form to today's instant transfer lettering, and including sheets of stipple dots, natural grains, and ruled and crossed lines in various thicknesses. Other methods used in the chromolithographic reproducing of images included stumping, air brushing, grating, sprinkling and splashing.

Just as books and magazines are printed today, tins were printed in batches rather than one at a time. In order to prepare a final printing

stone which carried a good number of images ready to print off a full tin plate sheet, rather than one at a time, a number of transfers had to be taken from the key and colour stones. These were 'patched' up together (like a large regular patchwork) and then re-transferred to large stones the size of the tin sheets. Trolleys conveyed these heavy stones to the printing room, where it often took as many as four men to lift them into the flat-bed presses.

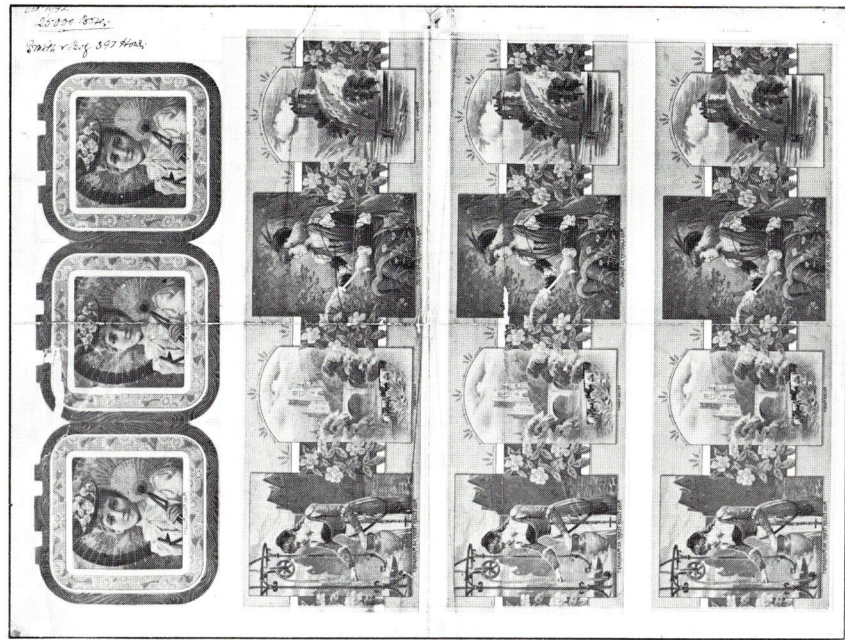

The final proof sheet for an 1892 decorative printed tin, pulled from the litho stones from which the tin sheets were to be printed. It shows how more than one tin was printed at one time by the 'retransferring' method.

If the tin plate sheets were to be printed by the transfer process (not to be confused with the transfer-reproducing method just described), sheets of paper transfers were printed up as in the normal chromolithographic process, except that the printing order of colours was reversed, ending up with a 'background white' which, when transferred to the tin plate, became the base coat.

With the offset printing process, up to six girls as well as the printer himself complete with long white apron and rolled up sleeves, were, according to Hudson Scott, needed to man each flat-bed machine: feeder, flyer, feeder of tapes, taker-off, hander-up and hanger-up. Not surprisingly the work was both hard and dangerous, and for this labour, in 1911 they received the princely sum of between 5s and 10s a week.

The tin plate sheets arrived in the printing room trimmed and, if necessary lacquered. A base white was printed first, sometimes up to two or three times to achieve greater opacity, but leaving bare the areas where the actual tin surface was to show through and making allowances for hinge and seam areas. It was imperative that the white was as good as possible since it compensated for the non-absorbent tin plate and was the basis on which the other colours (ground usually in the company's own colour mills) would be printed. When this was done the individual colours could be printed. Reduced colours

Chromolithographed proof for a decorative mustard tin issued by Colman's.

The tin box assembly room at Barringer, Wallis & Manners. As can be seen, many kinds of machinery were required for the construction of printed tins.

At Colman's decorated mustard containers were filled, sealed and then packed in paper wrappers printed in one colour with the same design as the tins.

(colours thinned down with lacquers) and tints were printed first to provide the foundation for the backgrounds, and to give depth of colour, thus allowing less ink but with a high hue concentration to be carried in the final printings.

The order of colours to be printed was as follows: buff, yellow, pink, light blue and light grey, all of which were tints. Then came red, dark blue, brown, the key black outline and finally a dark grey for tones and shadows. Lacquers and gold sheens could be applied at any convenient time during printing. If ten colours or more were used, proofing alone could take several days.

After the printing of each separate colour, the plates were stoved in ovens to bake the inks. A lengthy and delicate operation, the plates were not allowed to bake too hard for fear of the inks losing the elasticity needed for the shaping, folding and embossing stages. From the stoving ovens newly printed sheets went to the varnish room where one or more of a vast range of varnishes (tinted, clear, smooth, rippled, eggshell, crystalline and so on) was applied to give brilliance to the colouring and to preserve the printed surfaces during fabrication.

At this stage the tins were still 'flats' on a full printed sheet, but stamping-out, tooling and assembly had to be done as quickly as possible while the inks still retained their elasticity. Stamping dies and tools for the box-making department were for the most part constructed in the tool room and, although attempts were made at standardization, many tins, especially novelties, required special tools invented for their construction. Particularly complicated jobs, done by hand, necessitated many different operations, as many as forty to fifty being involved in the construction of a clock and two vases set at Hudson Scott. Female labour was used in the box-making departments as well as in the printing works, and again it was 'piece-work', women striving to approach the daily records (up to 18,000 at Hudson Scott) for notching and bottom-folding.

When the tins were finally complete, they were packed and delivered in bulk to the customers — the product manufacturers — by lorry, train, or, in the case of Hudson Scott's local deliveries to Carr's, by pony and float. On arrival at the customers' works, they were filled with the appropriate commodity, and sealed. At Colman's the decorated tins were packed in protective paper wrappers printed with the same design as appeared on the tin itself. Crated once again, the tins were now ready to speed on their way — perhaps, in Colman's case, in their superb brightly coloured liveried goods trains — towards destinations all over the world.

Sadly, few of those people actively involved in the production of decorative printed tins earlier this century are still about for consultation. Even fewer committed their working experiences to paper for the interest of future generations. Consequently information on many of the impressive traditional tin printing processes may soon be lost forever. Ironically it is only now, when it is virtually too late, that, hand-in-hand with the nostalgic interest in printed tins and other areas of old advertising art, comes a wide and genuine revival of enthusiasm for the fascinating techniques used in the manufacture of superb advertising items, such as those featured in this book.

Huntley & Palmers

The business of Huntley & Palmers was founded in 1822 by Joseph Huntley at a small bakery in Reading. For the next century the company consistently led the field in biscuit production, revolutionizing the industry and pioneering the use of printed tins for packaging provisions.

The bakery opposite the Crown Hotel, Reading, gave Huntley a distinct advantage over competitors, for this posting inn was on the main London-to-Bath coach route. Despite some eminent customers, the inn food was unpalatable and over-priced. Huntley was quick to capitalize on this; a delivery boy went daily to the inn yard to supply passengers while they waited for the horses to be changed. Such was the travellers' satisfaction that orders from grocers all over Britain poured into the Reading shop. With this expansion of the business, packaging to keep the biscuits fresh and unbroken in transit and storage began to become a problem. Huntley's solution was to use tin boxes, which his son, whose ironmongery shop was conveniently across the road, was prevailed upon to manufacture.

In 1841 Huntley was joined at the bakery by a fellow Quaker, George Palmer, under whose guidance the firm grew and prospered. Their name soon became a household word (the phrase 'take the Huntley & Palmers' instead of 'take the biscuit' was even used), and their reputation worldwide. 'Seldom a ship sails from England that does not bear within its ribs a Reading biscuit', they proclaimed in 1874. Their biscuits certainly accompanied the explorer Stanley through more than one African jungle, and were served at the King of Madagascar's coronation; black-market Reading biscuits even greeted the British expedition to Tibet, until then forbidden to foreigners. Pirates raided biscuit-carrying merchant ships, a landing party at Robinson Crusoe's remote Pacific island found only an empty Huntley & Palmers tin, and Prince Henry of Battenberg's body was returned to England in a rum-filled tank made of Reading biscuit tins.

Joseph Huntley junior, now joined by the Quakers Joseph Boorne and Samuel Beavan Stevens, continued to manufacture tin boxes.

To begin with, the tins carried paper labels, from 1851 often featuring the famous 'Garter and Buckle' trademark designed by White & Pike, Quaker printers from Birmingham (see page 6). But the paper labels were easily torn or lost, so alternative labelling methods were urgently sought. It was, however, almost forty years after Huntley & Palmers first used tin boxes that their earliest decorative printed tin was commissioned from the noted Victorian designer, Owen Jones, transfer-printed by Ben George (see page 13) and issued in celebration of their newly granted official royal patronage in 1868. Although fine results were frequently achieved with transfer printing, it was a cumbersome and lengthy method. In 1879 Huntley, Boorne & Stevens were exclusively licensed to print 'directly' onto tin by offset lithography, an invention that revolutionized all areas of litho-printing. Huntley & Palmers were quick to take advantage of their newly acquired expertise, ordering from them a selection of beautifully printed and assembled decorative tins to be put on offer for Christmas each year.

From the simple rectangular boxes of the early years evolved a series of progressively more adventurous 'fancy' shaped tins. Their surfaces were covered with scenes from British life, the Empire, topical events and fashions, even their bases were decorated, and the basic geometric forms were twisted and turned into all manner of shapes.

Although these 'shaped' tins always retained a place on the Chrismas lists, by the end of the century, with increasingly advanced tin box assembly methods, 'novelty' tins resembling specific objects began to creep into the catalogues. Initially ideas were relatively simple: handbags, trunks and other scaled-down versions of orthodox containers. In response to their popularity, designs soon became more adventurous — the more complicated the idea for a tin, the more the public seemed to like it, despite in some cases the almost total inaccessibility of the biscuits.

Interrupted only by the wartime halt in production, Huntley & Palmers issued these tins until World War II. Each year saw the introduction of new ideas, though popular tins were often re-issued, several years running. The most successful were produced in great numbers, but are not necessarily more commonly found today, especially since many featured delicate moving parts which were particularly susceptible to the attentions of children.

It is often difficult to date printed tins accurately, but luckily for collectors Huntley & Palmers had the foresight to keep a record of their very fine containers in the form of beautifully printed decorative catalogues (see page 32), from which much of the factual material in this chapter comes.

HUNTLEY & PALMERS

Most of the impressively constructed pictorial 'shaped' tins shown here were produced for Huntley & Palmers by Huntley, Boorne & Stevens from the mid-1880s on, when tin box making and printing of this kind was at its height, and they show the extents to which Huntley & Palmers were prepared to go in the interests of quality. Up to twelve printings, and several intermediary lacquer tints and varnishings to give the colour richness and depth were required for printing illustrations of this complexity, and enormous skill was needed to register them accurately with the chosen shapes. Complicated hand and machine tools used for stamping out and assembling the tins have long since been scrapped and the cost of equipping a workshop and employing labour to create tins similar to these would nowadays be prohibitive.

The picture on the left gives some idea of the variety of shapes issued for Christmas in their thousands, while that on the right shows geometrically shaped containers produced during the 1890s, examples of what must surely be the highest quality of illustration and construction ever achieved in tin manufacture. As shown here, the basic shapes can be found with up to four sets of different pictures: this is the case with the clover-leaf shaped *Algerian* (1894) in the foreground, which became *Mexican* in 1895. Highly detailed and descriptive illustrations provided a wealth of information for purchasers at home and abroad, not only on exotic foreign lands but also on many different aspects of British life and history. *Fire Brigade*, 15cm(6in) high, and *Universal (Athletic)* came out in 1892, *Hunting* and *Cavalry* in 1894, and *Nautical* in 1895.

Quality is apparent again in a feature that many collectors consider one of the most attractive of some Huntley & Palmers tins: their bases. Not all their tins have these elaborate black-on-gold printed undersides, and the reasoning behind the choice of tins to be treated thus is not clear. Perhaps it was thought to compensate for the lack of advertising on the tins' visible surfaces, although many whose construction and age might have deemed them suitable recipients bear the usual impressed tradestamp only.

Though they were by no means the only company to print tin bases in this fashion (see page 63), Huntley & Palmers were certainly the most prolific, and they continued to produce such designs, citing their awards and royal patronage until the outbreak of the Great War.

Both the shaped tins shown on the previous pages and many of the fine 'novelty' tins (those resembling specific objects) that were issued by Huntley & Palmers reflect artistic influences, although a noticeable time lapse often took place between a style becoming fashionable and its eventual appearance on tin. Consequently some amusingly strange stylistic combinations tend to occur. *Mirror*, measuring 16.5cm(6½in) long, issued in 1914, is an odd mixture of two influences. The figures in the illustrations are definitely of classical inspiration, particularly in their draped clothing and their braided hairstyles, while the lid's embossed copper surround is Art Nouveau in feeling.

Like quite a few other Huntley & Palmers tins, such as *Creel*, *Inkstand* and *Work Basket*, this tin's appeal was increased by the fact that it was designed to serve an extremely useful purpose after its original contents had been consumed. A curious departure from the wholly printed tin, its reflective mirror top is backed with a metal handle, which, by use of a swivel catch, can either by placed on a dressing-table or hand-held, as demonstrated by the reclining beauties on the tin.

Dragon and *Screen* are two of the many orientally inspired tins issued by Huntley & Palmers over the years. British interest in the East had been considerable since the 16th century when trade with China commenced. This fascination with the Orient continued throughout the following centuries, becoming even more marked after 1862 when the first public showing of Japanese art (after the long period of Japanese isolation) was held at the International Exhibition in London. Public interest was increased further when Gilbert and Sullivan's 'The Mikado' was first shown in 1885. Not surprisingly, oriental subjects gained, and retained, popularity with the designers of tin packaging, providing them with constant sources of inspiration for new containers, including the two shown here.

A superbly embossed golden Chinese dragon, traditional symbol of supernatural powers and the authority of the Emperor, writhes atop a lacquered green and brown 'strong flat rectangular tin of Chinese design called the *Dragon* tin', in which Huntley & Palmers offered 'separately our Cinderella, Swiss, and Veronique biscuits, and our Reception assortment. All 10/- per dozen. Suitable for afternoon tea etc. With calendar.' The tin was on offer in 1907/8 and measures 16.5cm(6½in) square. *Screen*, at 10s 6d, came out in both 1913 and 1914, and represents a full-colour folding Japanese screen. The scenes depicted on it show the influence of Japanese colour prints.

It is supposed to be impolite to ask a lady's age, which is just as well as Huntley & Palmers have no record of the date of this smiling plaster salesgirl, although her hairstyle indicates that she appeared in the mid-1920s. The interesting selection of novelty tins surrounding her are, from left to right: *Pillar Box* (1906-8), *Lantern* (1911-13), *Toby Jug* (also 1911-13), the most realistically scalloped *Shell* (1912-13), *Stories,* including Kipling's 'Jungle Book' (1911), *Stationery* (so popular that it was on offer from 1908 to 1914) and the earlier *Christmas Casket* (1903). Both *Lantern* and *Toby Jug* sold for 8s per dozen, while the 19.3cm($7\frac{3}{8}$in) long *Shell* was rather more expensive at 15s 6d per dozen.

The pair of tins the saleslady holds on her display plates are probably her closest contemporaries. *Egyptian Urn* was brought out in 1924 in celebration of Howard Carter's notable archaeological finds in Egypt, including the discovery of the tomb of Tutankhamun in November 1922. Classic examples of the popularization of a new fashionable style, which in this case can only be described as Hollywood Egyptian, the urns are decorated with embossed relief scenes and stylized geometric papyrus plants and birds.

More finely printed examples of Huntley & Palmers' continuing use of oriental subjects in novelty tin packaging. This insertion in their Christmas catalogue of 1928 described the tin: 'Chinese Vase. A copy of a centuries-old Chinese Jar. Packed with Royalty mixed at 26/- per dozen'. The tins stand 26.5cm (10in) high, and many people's initial reaction is that a 'top' of some description is missing. However this is not so, and, as can just be seen in the photograph, the tin breaks at the shoulder, the neck portion, though hollow, containing no biscuits. Huntley & Palmers enjoyed considerable success with this most accurate rendition of a porcelain jar, so much so that in 1934 they issued *Worcester* in an exactly similar shape, substituting a traditional Worcester Royal Porcelain Company pattern for the oriental look.

DECORATIVE PRINTED TINS

An interesting feature common to these five figural tins issued by Huntley & Palmers is that they all have moving parts. *Cannon* is the earliest of the group, and was on offer in 1914, costing comparatively little at 5s 6d per dozen. Mounted on moving wheels, it commemorates the Battle of Trafalgar in 1805, the roundels on its finely printed wood-grain sides 'bearing past and present views of Nelson's flagship "Victory"'. The sails of the wooden and brick *Windmill* (1924-7) turn, of course, and the woman standing in the doorway appropriately holds a paper-labelled tin of Huntley & Palmers biscuits.

Grandfather Clock, with its delicate clock face and moving hands, stands 29cm(11½in) high. Described as 'a lacquer tin packed with Wine Mixed Biscuits' it sold from 1929 to 1932 for 2s, although that was fairly expensive compared with the 1930 *Perambulator* which cost 4d less. The detail of the drawing on this tin, and the printing of the chuckling child and teddy tucked under the eiderdown are superb, as is the tin's construction. Compared with the pram, *Farmhouse* (1931-6) is fairly simply constructed, despite the protruding fence which could be neatly tucked away by folding it round to the back.

The wealth and realism of descriptive detail — the cows being milked in the dairy, the waterbutt and the wisteria creeping up the wall — make this tin remarkable.

With the increasing popularity of novelty tin packaging, and particularly tins such as *Windmill* and *Perambulator* which had moving parts, fresh complications arose: how to ensure both safe transportation and safe storage of tins in warehouses and shops. A new system was needed to cope with protruding appendages and irregular shapes. Imagine the look of total disenchantment on young William's face had he discovered that his fine 23cm(9¼in) high *Windmill* had lost a sail somewhere between Reading and London. Consequently a rather ludicrous situation arose whereby cardboard packaging was introduced to package the tin packaging of the product. Mass-produced, these cartons were cut and folded from single sheets, and illustrations of the tins contained within were printed on the brown cardboard. Few have survived the years; we can guess how long William's lasted in his haste to get at the gleaming toy and biscuits contained inside it.

Can there really be five faces to this 17.5cm (7in) high rectangular tin? The answer to the question is complicated but interesting. In 1909 Huntley & Palmers issued a splendid brightly coloured and embossed tin. It depicted on each of its sides a delightfully caricatured infantryman on sentry duty from each of the English, French, Russian and German armies. The colours or flags of the respective nations were also shown, and the soldiers themselves extremely well portrayed, with much attention given to national characteristics, such as the different moustaches, and to the accuracy of the different uniforms.

Although *Sentry Box* was an extremely successful tin that year, it was for some reason not re-issued the next. In fact it was four years before it was resurrected, a lengthy time lapse in the consumer packaging industry, which would normally have capitalized on the success of a particular tin by continuing its issue in successive years. However Huntley & Palmers did re-issue *Sentries,* now filled with Household biscuits, selling at 5s 6d per dozen, in 1913. Note the subtle name change and, much more important, the substitution of the earlier, friendly-faced German guard with a now more politically compatible Belgian. Solution to the riddle? This was a patriotic gesture from Reading during a time of political and military turbulence, when the King's cousin, the Kaiser, had commenced the preliminary rumblings that heralded the Great War.

Collectors, whilst agreeing on the quality and desirability of both tins, disagree as to which is the rarer. In fact, neither is particularly common, but it is interesting to note that several of the earlier version have been found with the German sentry defaced.

DECORATIVE PRINTED TINS

The bizarre origin of some of the ideas for new novelty tins is well illustrated by the strange tale of *Marble,* issued by Huntley & Palmers in 1909.

Huntley, Boorne & Stevens, tin box makers and printers of Reading, had since the earliest days supplied much of the packaging for Huntley & Palmers' products, and in particular the special issues of decorative tins produced every Christmas. This particular year, as was customary, mock-ups (made of wood with rough illustrations pasted onto them) of the proposed new Christmas tins were prepared, and presented to one of the Huntley & Palmers directors who had come to view, and hopefully approve, the new designs. A vase was put forward as the main tin and, in order to create an even more impressive presentation, it was displayed on a rough imitation of a marble plinth. The story goes that the director took one look at the vase, knocked it flying and chose the plinth instead, immediately seeing the potential in what proved to be one of Huntley & Palmers' most popular Christmas tins.

The success of *Marble* resulted in the issue of *Statuary* the following year, 1910, an exactly similar tin but with the colouring — green background and brown niches — reversed. Both tins measure 18.5cm(7¼in) high.

A complicated roll-top mechanism is the unique and distinguishing feature of the very decorative *Casket (Sèvres).* Because of this the tin is frequently but incorrectly referred to as 'Roll Top Desk', since the two sliding sides of its lid function in exactly the same way as the 'S'-shaped roll-top of the familiar 19th-century desk.

The tin was issued in 1909-10, following the success of the identically shaped *Casket (Fame)* which had appeared in 1906-7 with different illustrations and colouring. The *Sèvres* tins are decorated in pastel shades, predominantly pale green, with gold rococo style scrolls and foliage, and, according to the catalogue, 'embellished with enamelled pastoral scenes' in the oval medallions. The Art Nouveau style handles are another amusing incidence of the combination of artistic styles. *Sèvres* was Huntley & Palmers' heaviest tin, and one of their most expensive, selling at 35s per dozen. The tins measure 23.5cm(9¼in) high.

Bags, baskets, trunks, cases and caskets — traditional carrying receptacles provided one of the most obvious and successful choices of subject-matter for characterization in tin.

Most biscuit manufacturers issued at least one variation on this theme, but both the quantity and quality of those put out by Huntley & Palmers are impressive. Embossed basketwork, simulated leather, snakeskin and ivory were so realistic that their creation pushed construction processes to their limits. So much attention was given to detail and faithful reproduction that all of the tins shown here, from the dark green *Moroccan Hamper* of 1894 to the three *Wallet* tins of 1915, could easily be mistaken for the real thing.

Making their debut in Huntley & Palmers' 1907 Christmas catalogue were two particularly realistic tins. The 'pigskin' *Field Glass Case*, also issued in 1908 and 1909, and selling at 10s 6d per dozen, free calendar included, came complete with slots through which a leather strap could be threaded. The *Creel*, measuring 19.5cm (7¾in) high, and more expensive at 26s per dozen with free pocket diary, was described by Huntley & Palmers as 'an excellent imitation of an osier basket as used by anglers. It will prove of practical utility and can be easily affixed to a bicycle.' Few biscuit-loving fishermen could have resisted the temptation of acquiring this invaluable piece of equipment at some time during the six consecutive years it was on offer (1907-12).

Inkstand, issued in 1928, was a particularly unusual tin in that it contained a free gift in no way connected with the biscuits themselves — a small white china inkwell. Even by inventive standards already set, this was clearly a tin with an exceptional utilitarian value. Unfortunately for collectors today, Huntley & Palmers chose to use on the tin not one of their resilient patent fasteners, but a metal clasp which became easily separated from the tin itself, as it is often missing from an otherwise perfect container. The lid features an oval scene of a fox hunt, flanked by hunting horns and foxtails, while inside Huntley & Palmers appropriately offered their customers a selection of 'Full Cry Biscuits'. The tin is 23.7cm (9⅜in) long.

DECORATIVE PRINTED TINS

No publication on tins would be complete without the celebrated Huntley & Palmers 'book' tins. Of all the incredible and inventive novelties they offered, these were some of the most popular tins with customers then, and they remain so with collectors today. Although other firms offered the occasional book tin, none explored the potential of this idea as Huntley & Palmers did. Between 1900, when *Library* came out, and 1924, the last year in which a new book tin appeared, no less than ten variations of the 'book' theme were issued. During that time over 650,000 book tins were issued by the company, which largely accounts for the fact that they are so well known today.

Shown here are examples of each of the different book tins, apart from *Dickens* which appears in the catalogue on page 32. They all exhibit the same superior quality of workmanship which characterizes Huntley & Palmers' packaging, and the reproduction is so realistic that they are often mistaken for the real thing. Of the two single books, the red 'leather' *Book Tin* appeared in 1920, based on a volume of 1704 in the British Museum. Customers who chose the brown version of the same name, the last tin in book shape to be issued in 1924 by the firm, were informed that it would 'always be useful as a sandwich tin'.

The fine *Literature* (two examples: centre, and back row), offered in 1901-2 and standing 15.8cm(6¼in) high, is the only tin which can boast marbled edges to its eight classics, including 'Pilgrim's Progress', 'Robinson Crusoe' and 'Pickwick Papers'. It is also unique in that the company's name is printed on the marbled edge rather than impressed onto the flat inside of the tab opening lid, as in the other sets. This tab is clearly visible on the earliest tin of the group, the dark red *Library* (1900-2), one of whose volumes has the ambiguous title 'History of Reading'. It sits atop two *Waverley* (1903-4) tins, representing some of Sir Walter Scott's famous historical novels. Other favourites are depicted in *Bookstand*, a superbly embossed model of a revolving bookcase, distinguished by being the only tin on the surface of which Huntley & Palmers printed the date of issue, 1905 (bottom left corner), and in *Books* (1909-10) placed between Georgian bookends, their spines acting as the lid.

Beautifully drawn and printed with decorated capitals and borders, the superb *Good King Wenceslas* (1913-14) is without any question the epitome of the Christmas biscuit tin, the hinge of its lid just visible along its central binding. Last and, in terms of size, least, is *Stories*, which contained a meagre ¼lb of biscuits when issued in 1911.

DECORATIVE PRINTED TINS

Like most manufacturers, Huntley & Palmers kept retailers and grocers well informed of their latest products by regularly distributing brochures of their wares. Each retailer would receive a catalogue, containing an up-to-date review of the full range of innumerable varieties of biscuits and cakes available. Accompanying this was a loosely inserted single sheet, attractively designed and brightly coloured, which offered their latest range of 'decorated' biscuit tins.

Although the Christmas market was the bull's eye of Huntley & Palmers' target, several catalogues were apparently distributed each year. Tins could thus be ordered throughout the year without the retailer having to carry the cost of holding large stocks, for, although the prices seem ludicrously cheap to us, these decorated tins were certainly luxury purchases. To save the retailer the embarrassment of having the customer know his mark-up (which was, however, controlled), Huntley & Palmers, who customarily printed wholesale prices alongside the brands, in this case issued the price lists separately.

These 26cm(10¼in) wide sheets are now extremely rare and much sought-after collector's items in their own right. With their charming decorative details and lettering picked out in gold and silver, they are just as much reflections of past times as the tins they feature.

These eight impressive tins span a quarter of a century of novelty biscuit tin packaging. *Universal* (1899-1900), the authentic trunk in the foreground, was one of Huntley & Palmers' earliest attempts at novelty design. Both subject and name indicate the international market at which the company aimed its products, and *Universal* was also probably the brand of biscuits offered in the tin — one of Huntley & Palmers' most popular varieties. Brass buckles, studs and leather straps are beautifully embossed, and each of the luggage labels bears the name of a different biscuit brand.

The edges of the realistically coloured stack of *Plates* (1906-7) are also skilfully embossed, as is the Art Nouveau motif on the snakeskin *Wallet* (1903-4), which has a brass clasp and handle. The lattice-windowed *Cabinet* (1911-12) contains a fine porcelain collection; it measures 18.8cm(7⅜in) high and sold for 15s 6d per dozen. The deceptively real *Hamper* (1904-5) has finely modelled handles and features Huntley & Palmers' name on its white tin tab — a splendid example of the inventive skill of the early tin box maker. *Bookstand* was issued in 1905-6, *Handbag* from 1904 to 1914, and *Fruit Basket* appeared in 1926.

Three most unusual and desirable tins. *Waterbottle*, described as 'a representation of a boy scout's waterbottle, filled with Household biscuits', came out in 1915, and was one of the few newly designed tins to be issued by Huntley & Palmers during the Great War. Production of decorative tins was cut drastically during the war years (only 200,000 were made for the home market in 1915, compared with 580,000 in 1913) and this no doubt accounts for the tin's rarity. Although tin plate was apparently not in very short supply, Huntley & Palmers were too involved in wartime activities — supplying Army biscuits and packing troops' rations — to be in a position to cope with the continuing demand for fancy biscuits and tins, especially since neither labour nor ingredients were readily available. As if to prove their support of the forces and in celebration of Britain's military might, Huntley & Palmers issued in 1922 the camouflaged *Tank* which trundled out of Reading on tiny tin wheels housed in the static tracks. A year later *Engine* appeared. A splendid, gleaming dark green puffer, it steamed ahead, size-wise, of all other Huntley & Palmers' competition, measuring 31.5cm (12½in) long with its boiler stoked to the footplate with Noël mixed biscuits.

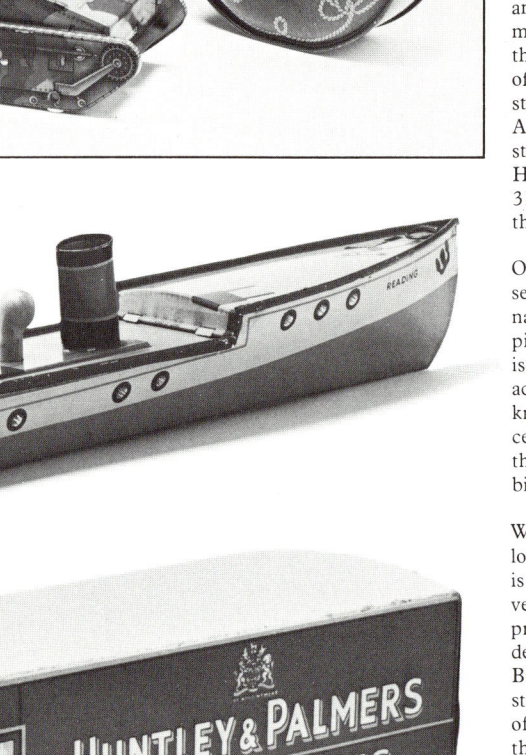

Over the years Huntley & Palmers offered several tins featuring pictures of ships or nautical themes, but this 43cm(17in) long pink-and-cream vessel (somewhat reminiscent of a River Thames tug) is the only tin actually in the shape of a boat that they are known to have produced. Issued in 1927, its centrally hinged lid opens from the bows, though goodness knows what shaped biscuits filled this unlikely looking tin.

What the contents of the 21.5cm (8½in) long royal blue *Tribrek* lorry (1937-40) were is also a matter for speculation. Its sides advertise both Ginger Nuts and a cereal product, Tribrek, somewhat unappetizingly described by Huntley & Palmers as 'The British breakfast food. Made to produce the strong and sturdy growth of the generation of tomorrow. Does not clog the teeth and thus renders them less likely to decay', but their 1938 catalogue has the tin filled with small shortcake. Unlike most vehicular tins, it is the back doors of this one which open, rather than the roof, and note also that the driver does not appear at the front window, only at the side one. One of the very last novelty tins to be produced before World War II, it sits on the page of Huntley & Palmers' 1940 catalogue amidst a humdrum selection of very conventional tins, signifying the end of the great years of novelty tin production.

BISCUIT TINS

BY 1870 HUNTLEY & PALMERS were the largest biscuit manufactory in the world, and for many subsequent years they led the field both in the production of high-class biscuits and cakes and in the packaging of their products in beautifully printed and constructed tins. But they were, of course, not without competitors whose produce, packed like Huntley & Palmers' in special decorated containers at Christmas time and in large paper-labelled storage tins for everyday use, also lined the shelves and counters of specialist grocers, provisions merchants and corner shops at home and abroad.

In business at the turn of the century were at least thirty rival firms all taking advantage of technical advances in production machinery (see page 44) and catering for people's rapidly changing eating habits. Demand for the new convenience food was so great that over 360 different types of biscuit were suggested by 'Law's Grocer's Manual' of 1896, so that the 'discriminating grocer' could keep his shelves 'well-stocked for the customers' individual choice'.

High-quality decorated Christmas tins were offered every year by the larger companies such as Crawford's, Macfarlane Lang and Jacob's, as well as many of the smaller biscuit firms. Unlike Huntley & Palmers, however, few of them produced many extravagantly 'shaped' containers featuring illustrations of different scenes, preferring to offer instead a wide range of marvellously inventive and original novelty tins based on real objects. Favourite themes included all types of motor vehicle, as well as trains, boats, globes and windmills, each one impressively constructed and embossed. Quite often, if a particular firm had a great success with a new tin, several tins based on the same theme would appear the next year and in a few cases some were even direct copies. However, over the years most companies developed their own recognizable styles, the emergence of which was often the result of a firm responding to an outstandingly successful tin by developing the same theme, as was the case with Macfarlane Lang's *Bird's Nest* and *Strawberry Basket*. Gray Dunn always produced tins with beautifully executed detailing on them while Jacob's, on the other hand, carried the art of the novelty tin to greater extremes than any other company with such incredible tins as *Sleigh* (page 40) and *Hobby Horse* (cover).

No doubt the middle-class market that these luxury tins were aimed at bought them more for the attractiveness of the container than for the biscuits contained inside, particularly those that were designed as toys or with a specific after-use in mind. Competition between the companies was most intense for orders for Christmas tins, which would eventually take pride of place on a Victorian parlour mantlepiece or dining-room sideboard, or become the favourite toy in an Edwardian nursery.

Since the orders from prospective individual customers and retailers were placed on receipt of the companies' Christmas catalogues, great store was placed on the quality of these sheets (see page 32). Hudson Scott, printers of both paper and tin, tell of the competition within the printing industry for the commission to print these illustrated lists. The amount of work involved in their production was enormous, from the depictions of the tins in gold, silver and full colour and the reproduction of, for example, the chocolate biscuits contained inside in exactly the right shade, to the struggle to keep delivery dates, thus ensuring that other companies' lists were not out first.

Printers also competed for orders to produce the containers themselves, but some manufacturing companies had made long-standing arrangements with particular printers, such as Huntley & Palmers with Huntley, Boorne & Stevens, or Carr's of Carlisle with Hudson Scott. In many cases containers were specially commissioned, but they could be chosen from the printers' annual illustrated catalogues of 'stock boxes'. These were tins held 'in stock' by the printer and available to any client, whose name could quickly be printed on the tin once an order had been placed, thus achieving a considerable time reduction in delivery of the order and a smaller financial outlay. Many small manufacturing companies chose to take advantage of this offer, which explains why different tradenames sometimes appear on otherwise identical tins. The names rarely appear on the printed surfaces of biscuit tins though there are, of course, exceptions. Instead they are more often discreetly printed inside the lids or on the undersides of the containers, or impressed into the base.

DECORATIVE PRINTED TINS

The range of novelty tins issued by biscuit manufacturers other than Huntley & Palmers was also enormous, and often equally as eccentric. Two rare and desirable novelty tins, made at Barringer, Wallis & Manners of Mansfield and offered by Macfarlane Lang & Co, are the impressively lifelike *Bird's Nest* (1908) and the mouth-watering *Basket of Stawberries* (1909). Apparently the clutch of blackbird's eggs was so successful that this rustic theme was continued with the issue the following year of the superbly embossed, and, as described in the catalogue, 'strikingly realistic depiction of a chip basket [punnet] of fresh ripe strawberries'.

Log Cabin from S. Henderson & Company came out in 1910, and the Far Famed Cake Company's charming vine- and chrysanthemum-filled *Greenhouse* and Hill & Company's *Punch and Judy Show* were probably issued some years before. Despite Mr Punch's aggressive behaviour, the show was, according to the tin, 'performed before royalty'. The children's market was still in mind in 1931 when Carr's of Carlisle ordered the ½lb jingling *Tambourine*, diameter 17cm(6¾in), from Barringer, Wallis & Manners' stock box catalogue. The same year, Paul Greville Hudson (1876-1966), for a long time a noted designer with Hudson Scott of Carlisle, produced *Accordion* (issued by Jacob's) which when opened and closed actually played, albeit only two notes on a primitive reed.

This golfer's stance and follow-through exemplify the superb quality of another rare and highly imaginative tin from Macfarlane Lang, *Golf Bag*, issued in 1913 and measuring 26.5cm(10½in) high. He cuts a dashing figure superbly embossed against a 'leather' golf bag complete with ball and tee pouch, the clubs doubling as the lid's handle. He is fully prepared for inclement British weather, from Harris tweeds to stockinged ankles, his equally decorous lady companion just one shot behind on the other side of the tin, as can be seen in the photograph on the cover.

As if her inimitable style could leave doubts in anyone's mind as to who designed these delightful Crawford's biscuit tin money boxes, Mabel Lucie Attwell's (1879-1964) name was featured on each one. The trio, *Bicky House*, 19cm(7½in) high (1933), *Fairy House* (1934) and *Fairy Tree* (1935), are original in both concept and design. The lower section of each contained the biscuits, while the upper part, separated by an ingenious tin ceiling, was the bank itself. Money could be retrieved from the bank by the removal of a tab located in the ceiling. Crawford's 1933 catalogue informed potential customers that *Bicky House* contained a 'bright selection of biscuits of certain appeal to children' and that it had been 'specially designed by Mabel Lucie Attwell, whose famous delineations of childlife in pictures and postcards have appealed both to parents and children for many years'. Certainly the influence of her distinctive and appealing style — chubby figures, creamy colouring and 'fat' lettering — is still seen today.

Two 1930s Macfarlane Lang tins which take popular fairytales as their inspiration: 'The House that Jack Built' and 'Goldilocks and the Three Bears'. Like the three Crawford's money box tins shown above, their style makes these tins distinctive, although it must be admitted that this is due partly to the strong influences they exhibit of rival manufacturers' products. Not only is the style of the illustrations very similar to that of Miss Attwell's work, but the 12.4cm(4⅞in) high *The House that Jack Built* is also remarkably close in both basic shape and date to Huntley & Palmers' 1931 *Farmhouse*, shown on page 26.

The same detailed illustration of the village blacksmith and his apprentice at work in the forge appears on both sides of *Anvil*, issued by Macfarlane Lang in 1910. Two children watch the scene from the doorway, while sparks fly as the smith hammers a horseshoe and his assistant tends the fire. The blacksmith's tools are also depicted on the front of the anvil, and a newly forged horseshoe lies atop the tin's lid. *Anvil* measures 16.5cm(6½in) high.

'Children are always interested in wild animals' was the bold assertion made by the tin box makers and printers, Barringer, Wallis & Manners of Mansfield, in the catalogue of stock boxes they sent out to their customers in 1907. This rather stark market analysis must have been essentially correct for from then on circus-inspired themes were frequently used for novelty tin designs. William Crawford & Sons of Edinburgh issued the *Menagerie Van* shown here in 1933 and it is extremely similar to that offered by Barringer's in 1907. Crawford's even acknowledged the debt to the earlier tin, describing theirs as 'a reproduction on the lines of a tin which had an enormous sale some years ago'. The Barringer tin was available for several consecutive years, and was still on offer in the 1913-14 catalogue.

Filled with ½lb 'Chocolate Home' biscuits, Crawford's wagon features beautifully drawn and lithographed circus animals behind bars: lions and a bear, as well as tigers and a zebra. The tin measures 12cm(4¾in) high.

Three late yet imaginative novelty biscuit tins for children from W & R Jacob & Co. The working model of a *Humming Top*, of which Jacob's catalogue says 'A strong and popular toy packed with attractively iced biscuits. A genuine novelty', was produced in 1928. *Trumpet* came out in 1934 and was promoted as a 'really well-made enamelled tin to arouse the interest of every child. Designed in cheerful colours and containing iced biscuits. Each trumpet in a heat-sealed container.' It even plays a somewhat nerve-jangling note. The yellow and green *Gypsy Caravan*, complete with folding steps, was issued in 1937, towards the end of the era of high-quality novelty tin packaging. Measuring 14cm(5½in) high, it proved such a winner that Chad Valley & Co, the famous toy manufacturers of Birmingham, subsequently issued it as a conventional toy.

This splendid 1930s motorcyclist looks at first glance like a tin toy, but in fact the uniformed rider has just picked up a ¼lb delivery from the Wright's Biscuits factory in South Shields, Durham, and is carrying them to their destination in his bright red and yellow sidecar. Admittedly the rendition of the subject is fairly crude, and the sidecar totally out of proportion with the motorbike and rider, but surely no novelty tin collector would ever complain. The tin measures 12cm(4¼in) high.

Two imaginative tins, complete with original cardboard 'outers', which must surely confirm that W & R Jacob's maintained over the years their place among the most creative and original biscuit tin packagers. *Sleigh*, issued from 1930 to 1932, is beautifully printed with crates of Jacob's own biscuits and a pair of snow shoes strapped to its top. For the benefit of their more youthful customers who had no suitably slippery surface available on which to use their new toy, Jacob's thoughtfully provided small wheels beneath the sleigh's runners. The illustration on the box is almost as finely drawn as the details on the sleigh itself.

The tin on the right of the picture was issued in 1936, a later example of the continuing inventiveness of the designers of biscuit tins. Let Jacob's themselves describe it: 'This graceful representation of the Coronation Coach provides a novelty of topical interest. Filled with iced animal and kindergarten biscuits. Each model packed in an attractive carton.' The 'topical interest' has always been thought to refer to the Coronation of George VI in 1937, but the fact that the description quoted here comes from Jacob's 1936 Christmas catalogue makes it almost certain that this tin was in fact originally issued to commemorate Edward VIII's accession in 1936.

The time needed to produce a novelty tin from concept to delivery obviously spanned some months. Jacob's must have decided shortly after Edward came to the throne in January 1936 to put out a commemorative tin in time for Christmas that year. When the king then abdicated on December 11th 1936, the company was no doubt relieved that, unlike many other commemorative items produced that year, their contribution did not have to be withdrawn, since the tin did not feature a portrait or name of any kind and was therefore equally appropriate for the new king. The tin measures 12.5cm(5in) high, and originally sold for 1s 6d.

'Everybody is asking for Macfarlane Lang & Co's biscuits', the company boasts on this realistic tin, partly no doubt because Macfarlane Lang & Co were using a range of Barringer, Wallis & Manners' very inventive novelty containers in which to package their ½lb selections.

Barringer's first offered *Telephone Box* as a stock tin in 1907, and its success was such that it was not withdrawn from their wholesale catalogue until 1914. The tin measures 22.5cm(8¾in) long, the front section containing the mouthpiece acts as the tin's lid, and the receiver is purely an accessory which can hang from a hook on the tin's side when not 'in use'. Rather ironically the wholesale price was 2¾d, considerably less than the cost of a local telephone call in Britain today.

The *Telephone Box* on the right is authentically reproduced, from its traditional pillar-box red paintwork to the embossed crown on each side. Issued by the Co-operative Wholesale Society of Crumpsall, Manchester, from 1873 pioneers of cheaper biscuits and provisions in general, the tin's restrained appearance makes an interesting comparison with the more decorative example from Mackintosh's shown on page 49.

Until quite recently it seemed that these 15cm(6in) tins might become the only reminders of the fast-disappearing original General Post Office telephone boxes. Demolition was however halted when the cost of replacing them became prohibitive.

A delightful Victorian scene captured in tin: Santa Claus has just visited this little boy and girl, *Our Darlings*, safely tucked up in their beautifully printed brass bedstead. Their stockings are filled with golliwogs, teddies and trumpets, while other Christmas presents, including the Victorian picture book 'Our Darlings', have been placed on the realistically rendered quilted eiderdown. A charming and rare tin, it measures 16.5cm(6½in) long, and was made by Barringer, Wallis & Manners for the Glasgow biscuit firm, Gray Dunn & Co Ltd.

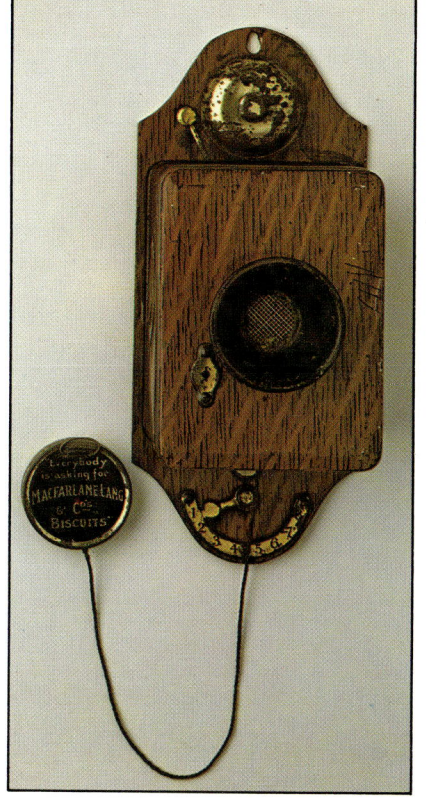

Somewhat surprisingly this multi-coloured tin is called *Bluebird*, referring not to the small bird found in North America in early spring, but to *Diomedia fuliginosa*, a species of albatross. This example was issued in 1911 by McVitie & Price, the Edinburgh firm founded in 1889 by Robert McVitie, a local baker, and C.E. Price, a Quaker who had previously worked at Cadbury's. The rather fiendish *Bluebird* was in fact issued by no less than four other companies at different times, including the Co-operative Wholesale Society in 1912. The 23cm(9¼in) high tin, styled after the grotesque earthenware birds produced by the Martin brothers of Southall, was certainly made at Barringer's, and was almost definitely one of their stock boxes since so many companies chose to issue it. The biscuits were retrieved by removing the bird's head.

The axe buried in *Yule Log* is in fact the handle of this seasonal tin, put out by Macfarlane Lang for Christmas 1910. They describe it in their catalogue as follows: 'A clever representation of a Yule log, with scenes illustrating the felling of the tree, the home-bringing, and the joyous Christmastide welcome.' The tin's diameter is 12.5cm(5in).

Household furniture was the inspiration of several novelty biscuit tins, and this 1932 *Chest of Drawers* (15.2cm(6in) high) is one of the better known. The particular shade of brown used by McVitie & Price on their reproduction is one not readily identifiable with any particular wood, though this is a minor quibble. The intricate 'inlay' and keyholes on this mid-Georgian commode are of course executed in 'gold'. The delicate feet are rather precariously attached, so care needs to be taken in handling the tin. Unlike other chest of drawers tins, it is the top of the chest which opens rather than the drawers themselves.

While McVitie & Price were not known for a prolific output of novelty tins, Macfarlane Lang, as we have seen, packaged their products in some of the most interesting and collectable containers. *Butterfly*, like many tins they offered, was made at Barringer's. Issued in 1901, at 16s per dozen, it measures 21cm(8¼in) across and was described as 'a new elegantly shaped tin, having two compartments with separate lids. Decorated in Louis XV style, filled with Fancy biscuits', made presumably at Macfarlane Lang's Edinburgh factory, opened in 1886 solely for the manufacture of fancy biscuits.

Castle was issued in 1923 by Peek Frean & Co, a business that had been founded in 1857 by James Peek, a retired tea merchant, and his nephew George Frean, a miller, in Bermondsey, London. Although they were joined in 1862 by John Carr, whose elder brother ran the rival firm, Carr's of Carlisle, it was not until 1870 that the major part of their business was given over to the production of fancy biscuits as opposed to ship's biscuits and aerated bread. They soon became Huntley & Palmers' most consistent competitors, vying not just for retail accounts, but for royal patronage and for prizes and medals at national and international exhibitions, finally joining forces with them (and W & R Jacob & Co) under the banner of Associated Biscuit Manufacturers.

Peek Frean's attitudes towards packaging, distribution and promotion were advanced compared with those of some of the more conservative companies, and it is therefore not surprising that it was they who developed a marketing idea seemingly overflowing with sales potential: that of issuing tins whose appeal would be increased by the possibility of adding other similar or, as in the case of *Castle*, tins identical to the original one.

The flexibility of the individual sections of this stone fortress, complete with flagpoles and crested doorways, no doubt facilitated many a child's schemes for protecting regiments of toy soldiers with imaginative use of any number of tins. The biscuits could be retrieved by lifting the crenellations off the towers, which stand 15.2cm(6$\frac{1}{16}$in)high.

An interesting group which shows the same stock tin bought up from a tin box makers and printers and issued by different biscuit firms. Ordinarily the only noticeable difference between stock tins of this kind would be in the individual product company's name, located either on the inside of the lid or on the tin's base.

The central tin shown here, however, bears neither a firm's trademark nor the depiction of the Cheese Sandwich Inn on its lower half, despite the fact that its latticed windows are extremely similar to those of the other two tins. These are also unusual in that their respective company names, Meredith & Drew and Wright & Sons, are actually shown on the printed surface of the tin, forming part of the design of the Saloon Bar panels, as well as being placed more discreetly inside the lid or on the base. The tins all measure 12cm(4$\frac{3}{4}$in) high, and predictably enough contained Cheese Sandwich Biscuits.

DECORATIVE PRINTED TINS

This splendid *Berengaria* liner tin from William Crawford & Sons of Edinburgh was modelled on the actual vessel built in Hamburg, Germany in 1912 by the Vulcan Ship Building Company for the Hamburg/American line. Had Crawford's tin been produced before 1921, novelty tin enthusiasts would be searching for the 'Imperator', the name of the ship until she was acquired by Cunard during that year.

That Crawford's chose to model their 1928 tin on her was probably the result of the continuous publicity she received on account of her exceptional size of 52,226 tons; they no doubt also felt that her classic lines would lend themselves well to their concept of novelty design, from the side ladders, which act as the tin's clasps, to the masts, which when extended bring this model liner's height to 14cm(5⅜in).

The accompanying trade card is an unusual example of a retail advertisement for a biscuit tin. Advertising of novelty tins was as a rule minimal, since the grocer would not only have had the companies' catalogues available for inspection, but his shelves would already have held a selection of the illustrated cardboard boxes showing the tins contained inside.

This rare and inventive novelty tin, complete with original box, was issued by Carr's of Carlisle. The company claimed that it was within their walls that the biscuit industry had been born. This was of course hotly contested by Huntley & Palmers, but the fact remains that as early as 1840 a machine for cutting and stamping biscuits had been developed through the combined talents of the Carrs and a member of Hudson Scott, tin printers of Carlisle, with whom Carr's were to have a long association.

The *Grace Darling Lifeboat*, named after the heroic daughter of the Farne Islands' lighthouse keeper, is one of the few novelty tins to have been offered by Carr's and, as they explain on the box, it is actually 'two toys in one. If you want a toy to float, from its carriage slides the boat.' The tin measures 25.5cm(10in) long.

A beau and his lady companion set off in this working model of a *Regency Coach* issued by William Crawford & Sons of Edinburgh. Like many other high-quality novelty tins, such as the *Berengaria* and the *Flying Scotsman* offered by Crawford's, the construction of the tin is fairly complicated, although the rendition in this case is fairly basic. However, in contrast to the flat printing of the coachwork, the figures in the window are finely portrayed in the classic chromo-lithographic stippling technique. The tin measures 15.5cm(6⅛in) high.

The London North Eastern Region's engine number 4472 *Flying Scotsman* must surely have been chosen as the subject for this exceptionally beautiful tin because it was, and still is, one of the best-known locomotives in the world. She was one of the 'A1' class of locomotives designed by Sir Nigel Gresley which first ran in 1923. This fine 40.5cm(16in) reproduction issued by William Crawford & Sons appeared on the market five years later in 1928.

In 1969 'Flying Scotsman' followed the same trail as many fine novelty tins and the huge locomotive was shipped to America. Unlike the biscuit containers, however, after a three-year tour of duty she returned to England, where she now resides in a private museum. Her escape from the scrapyard was no doubt due to her international fame, though 'Flying Scotsman' probably owes some of this to the confusion that exists between herself and the 'Flying Scotsman Express' which, since June 1862, has left Kings Cross Station, London at 10.00 am every weekday on its journey to Edinburgh.

Looking remarkably like an early clockwork model engine, the impressively embossed wheels and piston rods of this beautifully modelled Crawford's tin are in fact non-functional, though the locomotive does run on a series of tiny discs located on the inside of the pressed tin surface. Although it is almost impossible to believe that this really is a biscuit tin, the biscuits were in fact stored in the main boiler, which opened at the front.

Although the Macfarlane Lang *Violin Case* has on occasion been said to bear a strong resemblance to a coffin, this in no way detracts from the tin's collectability. Issued in 1903, it was described in their catalogue as an 'artistically decorated representation of an inlaid maplewood violin case filled with special assortment. 10/6 per dozen.' It is an unusual tin in that the catch and handle are made of brass rather than the more usual pressed tin.

The same company had in 1900 issued the comparatively large embossed *Book* tin, 21.5cm(8½in) high, with their name printed in black on the inside of the lid. The decoration of the tin, again according to the company's catalogue, was 'in facsimile, slightly reduced in scale, of an old French binding in the middle of the 16th century in the Gothic style.' The original volume had apparently been purchased on behalf of the nation in 1846, and is now in the British Museum, London.

The *Lighthouse*, a much later tin, is the only example in this book of a tin issued by the Hughes Biscuit Company. Made at the Barnsley Canister Company, it features an unusual retractable dispenser with four apertures (the top of the lighthouse), fully extended in this photograph, from which the biscuits could have been individually shaken.

The embossing on the red, yellow and green biscuit tin *Motor Bus* from the Co-operative Wholesale Society of Crumpsall, Manchester, is of the highest order, right down to the driver's quilted seat. The unseen sides of the bus show a different collection of passengers and the bus conductor blowing his whistle, but by far the most intriguing traveller is the top-hatted gentleman reading his newspaper at the front of the bus. Some people think he bears more than a passing resemblance to Edward VII, who was reigning when this tin appeared in Barringer, Wallis & Manners' stock box catalogue of 1907. 'No. 1.' in that year's 'Toy Series', according to the catalogue, the tin came in a cardboard outer, contained ½lb biscuits and measured 19cm(7½in) long.

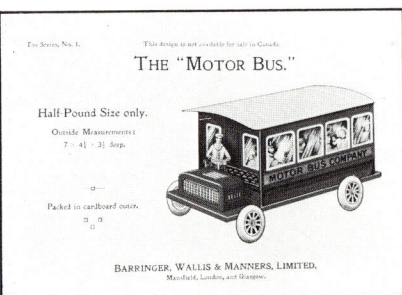

CONFECTIONERY

MANY OF THE CONFECTIONERY companies, particularly Rowntree's, Mackintosh's and Sharp's, also chose to package their products in tin for the same reasons — freshness and protection of the goods — as the biscuit companies did, and the range of tin containers the confectionary came in was as wide as the variety of sweets, toffees and chocolate they offered. Along with giant storage bins came simply shaped tins with bright illustrations, beautifully detailed miniatures and inventive novelties reminiscent of those issued by the biscuit manufacturers.

Because children were traditionally the market at which these products were aimed, much of the design of novelty sweet packaging was geared towards the interests of the contemporary child. Many tins consequently took the shape of traditional toys such as cars and buses (see page 84) or seaside pails and buckets, while in later years others took advantage of popular childhood enthusiasms of the day such as Felix the Cat (page 54). Several companies developed specialities for which they became particularly well-known: Sharp's parrot cages and Clarnico's and Rowntree's delightful miniature tins probably intended for Christmas stockings.

Not all confectionery companies issued novelty tins with children solely in mind. Fryer's of Nelson (Lancashire), whose product — medicated Victory-V cough sweets and lozenges — was very specialized in comparison with other companies', hold the distinction of being the first confectionery firm to use tin for packaging. Issued annually from before the turn of the century, many of their superb tins, containing a season's worth of sweets, were intended to appeal just as much to adults as to children. Some, such as the clock sets made at Hudson Scott and Barringer, Wallis & Manners, or the kitchen scales, were specifically designed for household after-use. As with the biscuit manufacturers, competition between Fryer's and their rivals to get their goods into the shops was fierce and on occasions pretty unethical, particularly since once a retailer had taken a 16lb tin or set of tins he would be fully stocked for the whole season.

A code of practice was therefore set up by Fryer's to which competing manufacturers were bound. On September 1st every year, the actual day and not a minute before, the firms' representatives would deliver samples of that year's novelty tins to the wholesaler, whose representatives would be waiting on the doorstep to receive them, preparatory to dashing out to get orders from the retailers.

The actual tins were packed in 'carrier' cases for the representatives, made in Fryer's own workshops and ready for delivery on that magic day — September 1st. The excitement was great and deliveries were carried out as fast as possible, just like a military operation. Everyone was honour-bound (officially) not to jump the gun in any way. Jack Haythornthwaite of Fryer's (see page 50) remembers in the twenties 'leaving Nelson at dawn when I was 17 in a Ford T-type lorry loaded with these cases, dropping them off starting at Doncaster before 6 a.m., then to Wakefield, Bradford, Shipley, Saltaire, Keighley and then back like mad to reload. The excitement was tense and skulduggery not far behind. It was not uncommon for a manufacturer's representative to come to his car and find his tyres flat; one poor fellow had water in his petrol and once there was a real punch-up in Wigan, Charles Davenport, our Lancashire representative, claiming damages for a smashed silk topper and a torn frock coat'.

With the exception of most of the novelties and a few early toffee tins, the real boom in packaging confectionery in tins occurred later than with many other products and chocolate in particular was always much more frequently packed in boxes decorated with beautiful paper labels. While examples from the First World War are scarce there was, however, a massive output of toffee tins during the late 1920s and 1930s. Every flavour of toffee seems to have been packed almost exclusively in tin, individually wrapped or in slabs with small hammers provided by the manufacturers with which the retailers could break off smaller portions for sticky-fingered young customers. The toffee hammers are now eagerly sought by collectors.

Toffee tins were comparatively simply constructed but featured bold, 'hard-sell' designs and lettering, quite often at the expense of the illustrative detail apparent on other contemporary tins. The relatively naïve scenes depicted on them are executed in a fairly basic colour range, printed in solid colour to give a strong brilliant effect (as can be seen on page 52), no doubt partly because they were intended as transitory containers appealing for the most part to children. However, while they tend to lack subtlety and detail, they are nonetheless very attractive nostalgic containers and, since they are quite commonly found, relatively inexpensive.

DECORATIVE PRINTED TINS

In common with many biscuit manufacturers, confectionery companies also chose to package their products in novelty containers, designing them to appeal particularly to the children's market, as these toffee tins illustrate.

The monocled, bowler-hatted gent, known as Sir Kreemy Knut, was the familiar trademark on Edward Sharp's 'Super-Kreem' toffee tins for over a decade. The idea of using such a dashing cartoon character as Sir Kreemy to promote toffee was humorously appropriate since he is portrayed as a well-to-do 'toff'. Though he was only immortalized by Sharp's in his two classic relaxed poses, feeding his pet parrot or lounging on his cane, he was kept busy in other spheres of promotion and even holds the dubious distinction of being the only character from advertising to have been used as a car mascot.

Both brightly coloured parrot cages can be hung up, as can the 14.5cm(5¾in) high *Lantern*, with its winter scenes and town crier, offered by J. Lyons & Co, who also issued the *A.1. Balloon Basket*.

Chocolate selections were rarely marketed in tins, though Rowntree's novel idea of the 1930s, shown on the picture's left, served its purpose very well. The 15cm(6in) long tin in the foreground opens in two. The lower half containing the chocolates has a small oval recess in its base and this is designed to slot into the matching protrusion on the top of the lid, thus creating an unusual and elegant form of presentation, as demonstrated by the retail confectioner's display example made, chocolates and all, from tin.

J.S. Fry & Sons, the Bristol confectionery firm, incidentally connected with Barclay & Fry, original patentees of offset lithography (see page 14), also presented a chocolate assortment in tin — this unusual life-size apple, designed for re-use as a string ball container.

The ingenuity of these tins was obviously not appreciated, judging by their fierce expressions, by the *Chiefs Thundercloud* and *Red Fire* of the Lyons tribe. Mackintosh's, on the other hand, showed an untypical lack of imagination in borrowing the idea of housing their products in the boiler of a tin *Flying Scotsman*. It is perhaps appropriate that their version was 15cm(6in) shorter than the splendid Crawford's locomotive on page 45, from which the inspiration was probably drawn.

CONFECTIONERY

Sadly Rowntree's of York have little information on these unique and beautifully produced miniature novelty tins, the largest of which, the *Grandfather Clock,* is only 9cm(3½in) high. Several advertise that they contained cachous, although it is more likely that those with larger openings held a sachet of cocoa. Several issues of these captivating give-aways were apparently offered but no complete list exists of the different shapes, so hitherto unknown examples do occasionally appear.

Amongst the most interesting are the *Punch and Judy Show,* the smallest tin to incorporate a moving part, a minute lever, which when activated causes Mr Punch to clout poor Toby the dog, and the exquisite *Christmas Pudding* shown in the photograph below. Complete with holly sprig on top, it opens, as you might expect, by taking the pudding off the plate. To give an idea of the pudding's size, another fine piece of early tin printing has been included. Fry's Victorian folding ruler features the famous 'Five Boys' chocolate trademark, frequently used in the company's advertising.

Both these pretty novelty toffee tins were made at Barringer, Wallis & Manners. As can be seen from the illustration of the sailor-suited child on the beach, Turnwright's *Toffee Bucket* was perfect for building sandcastles after the contents had been consumed. Unlike the example shown on page 41, the 15cm(6in) high Mackintosh's *Telephone Kiosk* bears a charming illustration of a little 1920s girl inside the open door searching for a number in her address book. The slot on the top of this money box is just long enough to take an old penny, and the 'telephone' bell rings when the handle on the side is turned.

49

Amongst the largest and most distinctive tin containers ever produced were those issued by the confectionery firm, Fryer & Co of Nelson, Lancashire, as retail dispensers of the somewhat unusual but very effective Victory-V medicated gums and lozenges. In about 1860 the Fryer family, founders of this 'cottage' industry, were bought out for £1000 by a Dr Smith of Bolton, who was already involved in producing medicated lozenges for the poor. His brother, W.C. 'Toffee' Smith, was installed to run the firm, and it was he, together with his successors, the Haythornthwaite family, who built the company into a nationally known business, partly by introducing the use of the superior and innovative packaging so admired by collectors today.

How the firm's products came to be known as Victory, and subsequently Victory-V, gums and lozenges is one of the many anecdotes associated with the company's history. Dr Smith and his brother had been seeking a trade name for their new business. Whilst walking to lunch in Bolton the doctor looked up and saw a mill chimney bearing the name 'Victory Spinning Mill'. 'There's your name, William! "Victory", the flagship of Lord Nelson — our town!' The design of the gums featured an impressed 'V', and customers were wont to say 'I'll have some V-gums', so J.W. Haythornthwaite snapped up the idea of alliterative advertising, promoting the product from then on as 'Victory-V'.

The company holds the distinction of being the first to pack sweets of any kind in tins, and their early printed containers, such as that depicting the muffled gentleman in a railway carriage (Victory-V for Cold Journeys!), are of the highest quality. One main novelty tin was issued each year in quantities that varied from 100,000 to 150,000. The subjects depicted ranged from the tins shown here to household scales, a Victory train, a tramcar, the Mayfair carriage clock, and several versions of the vase and clock sets. Unlike the biscuit manufacturers, no emphasis was placed on the Christmas market, since these outsize tins were never intended to be sold direct to the public, but were aimed at inducing the retailer to buy, and to buy in quantity. Competition between manufacturers was of course fierce for if a tin containing 16lb of sweets was sold the retailer would be full for the season. He probably never realized that he had actually paid for the tin in which he bought his stock — about 2d extra per lb on gums and lozenges contained in any of the novelty tins, but if he managed to sell them when empty to customers he could make a small extra profit.

Shown here are some of the most impressive

CONFECTIONERY

of all the Victory-V tins. Both the Victorian basketwork *Cradle* and the hexagonal *Lodge*, 53.8cm(21¼in) high, were produced before 1899 and are of the same ample proportions as most of the company's novelty tins. 'The Hand that Rocks the Cradle, Rules the World' was Fryer's selling slogan for the beautifully printed rocking cradle, which also bears another fine feature common to several early Victory tins. In a style similar to some biscuit and mustard bases, the company had its name elaborately printed in black on gold on the insides of the tins' lids. Pull back the cradle's curtained section onto the quilted eiderdown and a minutely detailed depiction of the factory is revealed. Likewise the gatehouse is opened by lifting its gabled roof to display the firm's name precisely marked on each of its six segments.

The splendid chauffeur-driven limousine *Motor Car* appeared in 1912-13 with pale blue bodywork (as here), while in both 1913-14 and 1924-5 it was in grey, though on each occasion 12lb lozenges or 14lb gums filled this 45.6cm(18in) long monster tin. It is an exceptionally well-modelled container, with the passengers and windows portrayed in considerable detail.

Both the *Victory Panorama* and the 25.3cm(10in) high *Church* were made at Barringer, Wallis & Manners. The rendition of the church is accurate, from the Gothic doorway to the tiled roof which acts as the tin's lid. The *Panorama* is in the style of a traditional proscenium-arched theatre and the panorama, a continuous passing scene running on rollers on either side of the stage, was moved along by turning the two handles on the tin's roof.

Probably the best known of all the Victory tins, however, were the clock sets. J.W. Haythornthwaite, who once said he'd bought more clocks than anyone else in the North of England, dreamt up the idea and several different variations of this exceptionally popular trio were produced between 1910 and 1925. While they differ in shape, colouring and design, they all retain the principle of the central working timepiece and two accompanying vases. It was not just the 33.6cm(13¼in) clock's conventional wind-up mechanism that was reached through the hinged door at the back of the case — it also housed 8 of the 16lb of gums or lozenges that made up the offer. The *Warrior* tins shown here are embossed in full colour against a rich brown background, and were most likely made at Barringer, Wallis & Manners, since they are included on the Barringer's splendid display lorry which is shown on page 14.

DECORATIVE PRINTED TINS

A varied collection of toffee tins which admirably illustrates changing styles of decoration over the years. The Victorian *Clarnico* tin and that depicting Manchester Town Hall, designed by Alfred Waterhouse, are early examples, while the Walter's *Fireside Assortment* was issued in the 1940s. The word toffee or toffy is supposedly a derivation of 'tough', but this was not an aspect of the sweet to which the various manufacturers chose to draw attention. Instead they went to the other extreme in their use of pretty, lighthearted and often sentimental illustrations.

Most makes were usually available in a large variety of sizes, depending on the intended uses of the different tins. The giant semi-cylindrical Co-operative Wholesale Society tin, *Dainty Bits*, was designed as a store bin, while the minute Yeatman's *Westward Ho* toffee drum, measuring just 4.5cm (1½in) high, sold for only 1d. This is a proudly patriotic tin, displaying the flags of Britain, France, Belgium and Russia, with a stern image of Lord Kitchener on its underside.

One particularly interesting retail bin shown here is the largest of the three *Palm Toffee* tins. Pierrette is shown taking a slab of toffee out of another Palm Toffee tin, on which she is also featured. This tin illustrates one of the two basic methods of marketing toffee. Slabs, moulded like bars of chocolate, were broken by the retailers with small hammers and then weighed into paper bags. Other selections, as shown on the *Dainty Bits* tin, were individually wrapped.

An impressive range of flavours were available, some of which can be seen listed on both the *Crusoe* tin, which has a space left on the front for a paper label to be pasted on, naming the variety inside, and the *Meadow Cream* tin. Crusoe makes 'a great discovery', the Sharp's little girl thinks 'it's alright', while Slade's describe *Merry Xmas* toffee as 'pure as the crystal spring'.

Two 10cm(4in) high toffee tins made at Barringer, Wallis & Manners.

Confectionery

Rustic reminders of times past in the English countryside, in the form of two Mackintosh's toffee tins. The traditional straw *Beehive* was produced in various sizes, this being the smallest. Though it does not qualify as a miniature, its dwarfish dimensions, 9 x 10cm($3\frac{1}{2}$ x 4in), held only $\frac{1}{4}$lb of toffees. Realistically detailed and coloured, the hive is encircled by buttercups, poppies and grasses of the hedgerow, bees buzzing all around it.

Keeping their distance from the bees, the windmill keeper and his dog, portrayed on the *Windmill*'s hidden side, patiently wait for the wind to come up. What they really need, however, is a ladder to reach the crankshaft, visible at the back of the tin, which turns the sails. Another favourite theme of the novelty tin designer, it is interesting to compare this version of the *Windmill*, with Huntley & Palmers' example on page 26.

At first glance this 7.6cm(3in) high *Box Camera* filled with Pascall's sweets looks just like the real thing — another example of the continuing enthusiasm for novelty containers which reproduced contemporary objects. The realism is increased by the accurate black and gold depiction of the camera fittings, and the addition of a functional carrying handle. The contents were reached by opening the 'portrait lens'. The tin's conceptual originality is, however, brought into question when it is compared with the 1913 Huntley & Palmers *Camera*, of which it is an almost exact copy.

Felix the Cat was the brainchild of Pat Sullivan, an Australian artist born in 1888, who was probably not aware that he was to bestow enduring feline associations upon the name Felix.

In addition to the cinema cartoons, such as are featured here, the mischievous tom reared his furry face in strip cartoons, on postcards and on sheet music ('Felix keeps on Walking' was his main claim to melodic fame), as well as in this rare example of an appearance on a toffee tin, issued by the R.K. Confectionery Company of Hull. As the strip on the tin's side shows, Felix has just removed a tin of Felix Cream Toffee from the confectioner's window and, dashing off with it, demolishes its entire contents, exclaiming 'Oh boy, what joy!' While R.K. Confectionery used Felix to promote their toffee, he in turn used their red, cream and black 15.2cm (6in) high tin to advertise his films, advising customers, 'If you like my toffee, see my films.'

GENERAL PROVISIONS

BOTH SHOPPING AND SHOPS changed dramatically in the Victorian age, particularly those businesses that sold groceries and household provisions. A whole range of new 'convenience' foods began to appear on the market and with the advent of better communications, particularly the railway (the 'Iron Horse'), supplies began to be widely distributed throughout the country. Prices dropped, wages were slightly better and tastes were changing enormously. Pre-packaged commodities, such as tea from the vast new plantations in India, formerly considered luxuries, were now readily available to everyone. No longer was the housewife dependent on limited local supplies, and cookery books such as Mrs Beeton's (right) now referred frequently to tins of this and packets of that in their lists of recipe ingredients.

In large towns, branches of multiples like the International and the Home and Colonial Stores opened, cutting costs by promoting their own brand foods, while corner shops and general grocers in villages supplied the local communities with lamp oil, candles, brooms and brushes as well as every kind of dry and fresh provision. Shelves and counters were piled high with a staggering range of brightly coloured packages boldly advertising the merits of their contents, which daily made their way into Victorian, and later Edwardian, larders and kitchens. Shop windows too, were crammed with packages and promotional material, and enamel and tin signs covered shop fronts already hung with other examples of their merchandise. Of the overwhelming range of products available, it has only been possible to show on the following pages a small proportion of those packaged in tin.

Although many of the companies such as Colman's and Lipton's, and brands like Cerebos and Bisto are still familiar today, a lot have of course sadly disappeared. Prominent in the list of companies who produced superbly printed tins but who are now no longer in existence is the Mazawattee Tea Company. Their intriguing tradename derived from a combination of the Hindi word 'Mazadhar' meaning 'luscious' and

the Singhalese word 'Wattee' meaning 'garden'. The fine tins they produced themselves at their New Cross (London) factory have done more than anything to keep the firm's name alive. The quality of their tin packaging was obviously much appreciated when the containers were originally produced, judging by the reference made in the May 22, 1908, issue of 'The Grocer' to a visit to Mazawattee by a party of French manufacturers, stating that 'much interest was manifested in the tin printing department where the beautiful designs appealed to the artistic taste of the French visitors'.

Apart from tea, the other major commodity which was traditionally associated with the British, and which also gained great popularity in Victorian times, was mustard. Of all the many different general provisions (other than biscuits) packaged in tin, only these two products were issued in special decorated containers designed to be kept rather than thrown away after the contents had been used. Although mustard was not consumed on such a vast scale as some things, its manufacturers were responsible for the issue of some of the most beautiful decorated Christmas tins ever produced. Quite similar in concept to the early 'shaped' tins from Huntley & Palmers and aimed at the same middle-class market, they were decorated with beautifully printed illustrations and bore only minimal advertising, the company name displayed more boldly on paper labels inside the lids. Each of the major companies' containers are easily recognizable by their different distinctive shapes and colour schemes.

Most of the general provisions tins, however, were made in simple utilitarian shapes and were hardly discreet in their attitudes towards advertising their contents. Featuring bold graphics and colours designed to catch the eye of the potential customer, many promoted products which are now redundant. From lighting tapers to blood salts, these tins are certainly evocative of the age in which they were manufactured. Perhaps more than any other printed containers, they reflect the changes in society over the last hundred years, both in technical advancement and in consumer attitudes.

DECORATIVE PRINTED TINS

For nearly a century every conceivable product, from blood salts to bifurcated rivets, was packaged in tin. Many were to be found crowded onto the shelves of the corner shop which in many areas, particularly country villages, had to act as ironmonger's, confectioner's and pharmacy as well as the grocery itself.

These two colourful groups of general provisions tins illustrate both the range of available products and the emphasis placed on eye-catching advertising packaging. Some of the products shown here, such as the *Silver Toned Radio Crystal,* necessary for the functioning of early cat's whisker wireless sets, have long since become obsolete and are unheard of today. Others, like *Slippery Elm* (below) from the Dr Thompson Pure Food Company, are still on sale. Containing more nourishment than porridge, and apparently the safest food for infants, this concoction combined 'the Nutritive and Emollient properties of Slippery Elm Bark (*Ulmus fulva*)', and a feeding chart was featured on the back of the tin. Still trading also are the largest British manufacturers of billiard tables, E.J. Riley of Accrington, who say that the same design of their tin was used for cue tips, chalk and spots alike.

Amongst the many decorative containers of the early 20th century is the beautifully printed *Quorn* custard powder tin (below), depicting the Leicestershire manufacturers' famous local foxhunt, the Quorn, in full cry around this 10cm(4in) tin.

Christmas speciality issues were not the only tins to feature the work of contemporary artists. The jolly cook on the Edwards' *Desiccated Soup* tin lid was painted by Louis Weierter in about 1900, and the lady, known as 'La Belle Chocolatière', featured on Dunn's *Chocolate Creams* came from a painting by Ratoire. She also appears on the American *Baker's Breakfast Cocoa* tins.

Ah Bisto! Will Owen's famous brother and sister have been sniffing the gravy for nearly sixty years now. Another artist whose work was immortalized on tin, Owen created this pair in 1919. Recently the Bisto (an anagram of the initials of 'Browns, Seasons, Thickens in One') Kids have been smartened up somewhat. Gone are the tattered shorts of a less affluent era, replaced with more fashionable jeans, but the basic image has continued to grace the company's packaging and advertising, one of the few trademarks to have remained in constant use since its creation. The large tin stands 25cm(10in) high.

GENERAL PROVISIONS

Decorative Printed Tins

GENERAL PROVISIONS

A fine collection of colourful string tins. These were a boon to grocers who, in the days before carrier bags, would parcel together customers' purchases and then secure the packages with binder twine. The idea of the tin was simple but effective. It was normally weighted with some sand in its base, though occasionally holes were punched through the bottom so that it could be screwed to the shop counter. The twine was pulled through a small hole in the lid, and some tins came equipped with a razor-blade attachment on which to cut the string. These tins were supplied free of charge to the retailer, who in return gave good value for his gift by prominently displaying the product name. Unlike the rest, however, the second tin from the top in the photograph bears no manufacturer's name, only an illustration of the types of twine that could be held inside. The tins' average height is 13.3cm(5¼in).

The Cadbury's *Cocoa Essence* string tin (inset) is a particularly fine example. It has an elaborate brass cutter, and bears on all sides the red and blue image that was used on much of the company's packaging and promotion.

Probably the only articulated lorry tin ever made, the container section of this brown and black 31.8cm(12½in) long delivery van can be detached from its driver's cab. The lorry's number plate makes it an easily datable tin — it was issued in 1932 by the International Stores, one of several grocery chains to realize the competitive market potential of their own brand name goods. It probably contained biscuits but could have been filled with another grocery, given the general nature of the company's business.

The 1920s housewife's dream — a complete kitchen set of four 13.3cm(5¼in) high tea caddies from J. Lyons (founded, incidentally, by the tobacco retailers, Salmon & Gluckstein). The Art Deco style tin on the left has a domestic thermometer attached to it, while the second tin, a two-toned brown container, includes a mirror and an appropriate motto, 'Time for Reflection', printed on the lid's inside. The calendar tin is, like the others, in shades of brown; the date is changed by moving the three serrated discs at the top and sides. And for those who demand their eggs boiled to perfection, a four-minute egg timer. The glass itself rotates, being attached at its centre to the inside of this pink and white tin.

The vast expansion of the tea industry in the mid-19th century was reflected in some beautiful tin packaging, since ready-mixed and pre-packed Indian teas now began to be distributed to grocers or direct to customers, as with the United Kingdom Tea Company tin shown here. Tea merchants to, amongst others, the House of Commons, they described their product as 'a right royal boon'.

Of this group the two tins featuring historical images are particularly interesting. The then Prince of Wales and Princess Alexandra, receiving a presentation of Terra Bona tea and coffee at Olympia on February 8th 1894, are depicted on the lid of the 15.2cm(6in) *Terra Bona* tea tin made at Hudson Scott, while *Triumph Tea* displays a portrait of General Sir William Booth, together with the crest of the Salvation Army of which he was the founder. The sentiments on the tin's back would doubtless have found favour with him: 'Remember that Triumph Teas abound in great strength and rich flavour, and make long friends wherever they are tried.'

The *Tabloid* tea tins, beautifully printed with the superbly Victorian image of a lace-cuffed hand complete with gold snake bracelet, contained a unique product — tea compressed into tablet form. 'Superior portability' was the claim made on behalf of this concentrate, popular with Victorian travellers. Each tin contained one or two hundred tablets, made by Burroughs, Wellcome & Company.

Miniature promotional tins were often issued by companies as samples. Of this group, displayed against a grocer's shop automaton of Edward VII and Queen Alexandra (who, when activated, raise their cups to their mouths and nod in approval), the best known are certainly the 3.75cm (1½in) Huntley & Palmers cubes.

Tiny printed tin versions of their paper-covered big brothers, they were described as 'Miniature tins packed with miniature biscuits — A replica of the foundation of a good biscuit stock.' Like other Huntley & Palmers tins, their seasonal appeal was strong: 'Very suitable for the Christmas stockings', and they sold at 21s per gross. First issued in 1922 with the traditional garter and buckle design, pictorial images were introduced in 1928.

The Co-operative Wholesale Society teapot is an intriguing novelty tin. It separates across its centre, with the handle acting as the hinge, and contains inside enough leaves to make just one cup of tea, as did the tea-collecting basket from Horniman's, which bears the strange comment 'Always good alike'.

The tea firm of Mazawattee was founded by a Mr Densham and his four sons, and although the company also manufactured chocolate and cocoa, it was the Ceylonese tea that they first introduced to the British public on 28th October 1878 that brought them international recognition. By the turn of the century they had without doubt become the foremost tea importers and distributors in Britain.

Over the years they packaged their products in a splendid range of printed tins. By far the most popular was the delightful *Old Folks at Home* series, depicting a bespectacled child drinking tea at the table with her bonneted and shawled grandmother, of which many variations in shape and size were issued. The original picture was painted during the 1870s, and depicted Mary Anne Clarke (the granny), the wife of George Clarke, bootmaker of Islington, and their neighbour's child, who stood in for the Clarke's granddaughter, Adelaide, who had unfortunately succumbed to an 'attack of nerves'. The scene has a very realistic, almost photographic, quality about it, from the characterization of the 'Old Folks' themselves to the fine china and lacy tablecloth. The tin on the left stands 16.5cm(6½in) high.

Fairytales were a popular source of subject material for tins at the turn of the century, as is shown by these three beautiful Victorian examples, again issued by Mazawattee. Little Red Riding Hood and the Wolf, here most unconvincingly disguised as her grandmother, was a recurring theme (the Champion's mustard tin on page 63 is another example), and the standard of the illustrations is often of as high quality as that found in many children's books of the same date.

The flat 19.5cm(7¾in) long tin in the front of this picture contained Mazawattee chocolates, while the others offered 1lb of their famous tea and these would have cost just 2s each. The tin in the centre shows five scenes from the Red Riding Hood story. In spite of her plight, she always remembers to keep a Mazawattee container prominently displayed at the top of her wicker basket. On the remaining tin, Red Riding Hood is reassured by the presence of some of her contemporaries: Little Bo Peep, Little Miss Muffet, Little Boy Blue and the Babes in the Wood.

Only in Victorian Britain could the marketing of such an inconsequential commodity as mustard have assumed proportions so great that the Norwich firm of J & J Colman, founded in 1823, was able to make a fortune out of 'what was left on the side of the plate'. This unparalleled success was in no small measure due to the enlightened structure of their organization, advanced for its day in many respects, from the calculated absorption of virtually all their competitors and the innovative staff welfare policies, down to the energetic advertising department set up in the late 1870s, which from that time consistently promoted the name of Colman's throughout the world.

Like the leading biscuit manufacturers, Huntley & Palmers, Colman's issued special 'decorated' mustard tins at Christmas, but they are mainly noted for the distinctive yellow, red and blue livery which systematically graced (and continues to grace) their product packaging — from tins, packets and crates to posters, tradecards, enamel signs and even their railway goods vans.

The 'outsize' tins shown here, the larger 51cm (16⅛in) high, boast the superior quality of their product, displaying every conceivable award, commendation and royal appointment. Tins like these often accompanied troops, explorers and expatriates to far corners of the world. A unique testimonial was sent to Colman's by Captain Robert F. Scott on his safe return from the National Antarctic Expedition of 1901-4: 'Sirs, I have much pleasure in informing you that the flour, mustard etc which you supplied to the "Discovery" proved to be of the best quality and in all respects satisfactory. Some portion which remained unconsumed on our return to England has been found to be in as good condition as when it was supplied. I must also congratulate you on the excellent way in which the goods were packed. This is an important consideration for an expedition, and in our case, no doubt accounted for the absence of deterioration.' A photograph of one of the giant tins, taken in the vicinity of the 'Discovery's' winter quarters, accompanied the letter.

The first recorded use of the Colman's Bull's Head trademark featured on both these tins and the brown 'copper' barrel shown below, was in 1855. This 5lb tin, 18.4cm (7¼in) high, issued in 1909, was accompanied by a note informing the customer that it would later be handy for storing tea, sugar etc. The blue-and-white 'Wedgwood' style tin came out ten years later in 1919.

Of these six late Victorian mustard tins, the five on the right were issued by Moss, Rimmington of Selby. The firm, which was taken over by Colman's in 1910, was responsible for some of the finest chromolithographic packaging, their tins distinguished both by finely drawn illustrations and also by the use of a distinctive colour range of blues and greens.

The bluey-green fairy tale *Blue Beard* tin (17.5cm(6$\frac{7}{8}$in) high; top centre) does not feature the usual varnishing and stippling techniques. In a style very similar to that used on children's book illustrations of the period, flat areas of matt colour fill the outlines of the images. Like most Moss, Rimmington tins, including the three smaller examples shown here (and some Huntley & Palmers containers), it bears on its base an elaborate black-on-gold trade stamp. Under it stands a particularly desirable 4lb tin, the sides and lid of which are adorned with famous turn-of-the-century actresses: Mary Anderson, Sarah Bernhardt, Miss Fortescue, Mme Marie Rose and Ellen Terry, each of whom is shown in portrait form and in a celebrated role. Yet another version of the Little Red Riding Hood story is depicted on the curving sides of the enchanting $\frac{1}{4}$lb Champion's tin on the left.

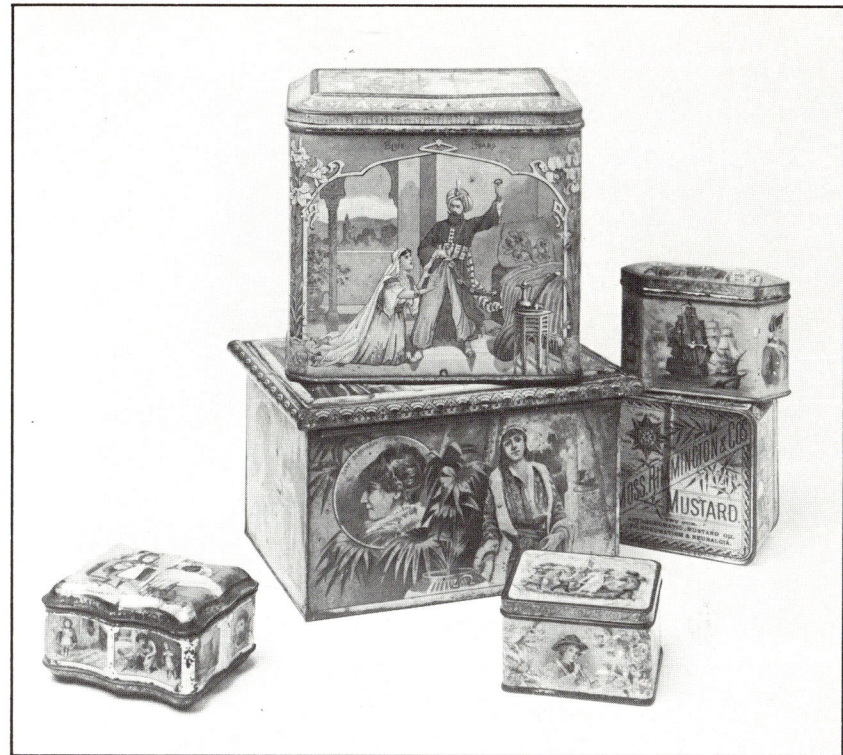

Mustard tins from Keen's and Colman's. Keen, Robinson, founded in 1742 at Garlick Hill, London and absorbed by Colman's in 1903, quote Shakespeare ('The Taming of the Shrew') on the lid of their 4lb yellow and black tin on the left: 'What say you to a piece of beef and mustard?' If the smiles of the ladies representing England, Scotland and Ireland on its sides are anything to go by, the answer was probably 'Yes, please'. The tin stands 19.5cm(7$\frac{3}{4}$in) high.

Colman's, who issued the tin on the right in 1906-8, was one of several manufacturers of the early 1900s to keep alive the memory of Grace Horsley Darling by using depictions of her courageous rescue on their containers. Born in 1815, she was the daughter of a lighthouse keeper on the Farne Islands. One stormy night, aged only 23, she put to sea in a small boat and went to the rescue of the shipwrecked crew of the 'Forfarshire'. News of her heroic deed spread quickly, and her untimely death of consumption just four years later set the seal on the legend of Grace Darling. *Heroes,* the name of this 16.5cm(6$\frac{1}{2}$in) high tin, also shows other less specific acts of bravery on its side panels.

Like biscuit companies, the leading mustard manufacturers also issued special tins for Christmas. Minimal advertising, if any, was displayed on the main surfaces so as not to detract from the tins' 'artistic merits'; instead the company name was either embossed on the base or threaded discreetly round the rim of the lid, which had a printed paper label affixed inside it. A sealed inner lid ensured that the powder maintained its newly ground freshness.

Keen, Robinson issued the two identically shaped tins in the centre, the lower of which, *The Rivals,* is named after Sheridan's play. The illustrations on the upper tin probably represent the Muses. All the other tins were offered by Colman's. Their high-principled attitudes towards superior packaging are exemplified by the upper right-hand tin, issued in about 1890, which made use of the talents of the distinguished artist, Sir Edwin Landseer, depicting on its panels four of his most famous paintings. A tin in a completely different style, *The Miller's Donkey,* 20.8cm(8¼in) long, tells the tale of Aesop's 'Fable', while *Big Game* (far left) first produced in 1899, was put out in 1901 in double quantities, following the withdrawal of *Sèvres,* which only got to proof stage since it was thought to be too artistic for British taste.

Seasons, which came out in 1899, must be one of the most desirable decorated containers ever issued by Colman's. Like several of their more progressive tins, including the withdrawn *Sèvres,* it was conceived and printed for them by the Parisian firm of Champenois. Amongst the artists under contract at that time to produce as much lithographic work as they required was the celebrated Art Nouveau artist Alphonse Mucha, whose first set of Seasons *panneaux decoratifs* appear on this tin. Despite the fame he achieved through his stage posters for the actress Sarah Bernhardt, he remained hampered by his time-consuming commitment to Champenois until his departure for America, although he did apparently remark: 'I was glad I was engaged on art for the people and not for the closed salon. It was cheap, within everyone's means and found its way into both well-to-do and poor families.' One of few printed containers to reflect absolutely contemporary artistic trends, only 30,000 of this decorative tin were issued, compared with Colman's usual order of 60,000. The tin measures 14.6cm(5¾in) high.

By the time they were acquired by Colman's in 1903, Keen, Robinson, reputedly the source of the well-known phrase 'As Keen as Mustard', had been producing fine decorated tins for over a quarter of a century. The lower of these two tins, both in the familiar Keen's 4lb 12.5cm(5in) long lozenge shape, shows scenes from the life of Admiral Horatio Nelson. The main sides of the tin depict, in heroic Victorian fashion, his progression from midshipman, while his naval decorations fill the smaller corner panels. *Dominions,* above, was issued in 1887 to commemorate the Golden Jubilee of Queen Victoria. It features on its sides views of the Empire, and on its lid medallions of the two monarchs whose reigns represented the extent of Keen's 145 years of production. As the company states atop the tin, its mustard was 'first manufactured during the reign of George II' in 1742.

Though no slot was provided in these *Azure Blue* tins, Colman's intended them for re-use as money boxes. Each tin originally contained 1lb of Azure Blue squares, a bleach-based substance to bring out the whiteness in laundry. The bottom two are printed in the standard No. 1 Azure livery, but the upper tin is in the bolder style more familiarly associated with Colman's starch products, and features on its sides medals such as the 1878 Legion of Honour. The tins measure 5cm(2in) high.

An occasional frustrating occurrence for the collector is the discovery that a choice acquisition bears no manufacturers' trademark. Both biscuit and mustard companies were culprits of this practice, but in the case of biscuit tins it was probably the result of stock tins (see page 35) not having been printed with the individual company names. These four Victorian mustard tins from Joseph Farrow & Company however, give a clue as to why this sometimes occurred with mustard containers. Loosely attached paper labels provided the trade information instead, and these tins are particularly unusual in that both external paper seals and inside lid labels have remained virtually intact, although in this case there is in fact also a minute Farrow's packet included in the illustration on one side. Some of the prettiest containers shown in this book, with beautifully drawn scenes from the nursery rhyme 'Who Killed Cock Robin?', these tins stand 7.5cm(3in) high.

For over half a century these blue, white and gold Cerebos salt tins, from their 'model factory in Greatham by the sea', were familiar sights in kitchens throughout the country. Anxious to avoid the potential consequences of a hastily salted meal, the firm provided their 3.25cm(5⅜in) high table-sized cylindrical tin with a patent pourer, the concept of which has not been improved on since.

The 28lb of salt contained in the largest tin shown here would have brought out the flavour in several hundred meals if the children of the house had not wasted too much of it in pursuit of chicks. The origin of this pictorial trademark is an old folk saying, that the way to catch a chicken is to put salt on its tail. The name Cerebos is a contraction of *ceres* and *bos* (Latin for wheat and ox, or, as in this context, meat). The product, according to an Institute of Hygiene notice in a 1906 'Illustrated London News', combined refined table salt with the phosphates of wheat and meat which are lost in flour refining and meat cooking, hence Cerebos, which 'restores these natural and valuable products in the daily food'. Although there is no visual reference to meat, it is now clear why ears of wheat are incorporated in the package design.

Most of the tins shown in this chapter would have been found in the grocer's shop, but a pharmacy would have been the location of this rare and interesting 12cm(4¾in) high talcum powder tin. Its originator was Louis Wain, one of the most prolific illustrators at the turn of the century, producing, according to one obituary, over 150,000 drawings. He was not averse to applying his skills to advertising and commercial art, but although it is known that he worked for Mazawattee (see page 93), this blue, white and gold container of about 1905 carries his only confirmed illustrations for a tin.

It is hardly surprising that the tin features a cat, since after 1884, when he acquired Peter, his cat, he was inspired to pursue and develop the cat depictions that were to become his immortal trademark. Always eccentric, Wain began to bestow human qualities on his pet, to the extent that, when his wife was dying of cancer, he expressed more concern over the cat, noting that Peter 'was the genius which gilded many a sorrowful hour, and lightened many a burden. To him properly belongs the foundation of my career.'

CIGARETTES AND TOBACCOS

THE NUMEROUS CIGARETTE and tobacco manufacturing companies in business before the turn of the century were among the pioneers of the use of printed tin as a commercial packaging material.

Packaging in tin provided distinct advantages to the manufacturers. Tobacco could be kept fresh in airtight tins and cigarettes remained uncrushed, however far they had to travel.

With the advent of new kinds of patent airtight tins, manufacturing processes extended a stage further than had previously been possible, since the variously flavoured blends could now be sliced or flaked before packaging. Until then, in an effort to maintain its freshness, tobacco was sold in a compressed form called a 'plug', the tobacconist slicing off the required amount with a plug-cutter, a kind of miniature guillotine. This traditional practice gave its name to many of the brands of tobacco now packed in tins, including 'Cut Plug', and 'Navy Cut'. When sailors purchased their duty-free allowance of leaf tobacco, they formed it into a roll by coiling string tightly around it. When they wanted a bit of tobacco, the string was unwound slightly and the pressed roll cut in slices, hence 'Navy Cut'.

The market at which these new brands were aimed was as wide as the range of packaging they came in. Unlike decorated mustard and biscuit tins, which were specifically designed for the middle classes, there was a cigarette tin or tobacco box to appeal to every type of potential customer, as is well illustrated on pages 68 and 73. Smoking cigarettes had gained great popularity with the upper classes and small exclusive firms catered for their requirements, offering hand-rolled and sometimes gold-tipped cigarettes in tasteful tins for the man about town, while at the other end of the market tobaccos and cheaper mass-produced cigarettes, such as Woodbines, had great success in working-class areas.

Competition between the different companies catering for all ends of the market was great, and the down-market tins were as well produced as the expensive ones, even though they were all only meant as transitory containers. Advertising in this business was particularly competitive, the actual tin being only one part of expensive campaigns to promote the product. Packaging designs needed to be striking and brand names memorable as tins piled high in the tobacconists' mahogany and glass showcases, or packed windows vied with one another for prominence.

In complete contrast to other fields of tin production where designers struggled to produce increasingly inventive and often outrageously shaped containers, the designers of tobacco packaging concentrated on beautiful images and flamboyant lettering. Although biscuit tin shapes became progressively more unrelated to their contents, tobacco and cigarette tins remained functional and suited to the product, and although many were eventually used for storage they were never intended as luxury objects or for specific after-use.

To compensate for the lack of complicated containers, the tobacco companies invented nostalgic and evocative names, imaginatively lettered and decoratively integrated with vivid illustrations depicting an extraordinary range of subjects, from lifeboat men braving storms to Red Indians pow-wowing outside teepees. Many represented healthy, outdoor pursuits or rustic country images, but the most popular theme, which retained throughout the years its foothold in the market, was that of naval heroism, depicted in many forms on such tins as *Lifeboat*, *Pilot Flake* and particularly Player's *Navy Cut*. More elegant alternatives which resorted to fantasy for their inspiration included *Tiger Lily* and *Night and Morning*, but, while many of the subjects bore no relation to the nature of the product, few companies went as far as that which issued the superb *Great Auk* cigarette tin, named after a flightless northern seabird extinct since 1840.

The same designs were often featured for many years on the different sizes of tins available, with only slight modifications. These details can sometimes help in dating tins and it is fascinating and intriguing to search for the tiny ship on the horizon one year, with no warning at all sunk without trace the next. The prices on tins are also useful in dating and on occasions it is possible to tell the date fairly accurately by the subject matter. The *Foursome Mixture* for example on page 68 must have been produced when the noted golfers Mitchell, Vardon, Duncan and Braid played together in the 1910s.

Since its formation in 1901 to present a unified front against James 'Buck' Duke (see page 75), the Imperial Tobacco Company has taken under its umbrella most of the tobacco manufacturers and retailers trading at the turn of the century, but many continue to trade under their own names. Other companies have totally disappeared, only to be remembered by some of the finest printed tins ever produced.

DECORATIVE PRINTED TINS

For over half a century Player's *Navy* cigarettes and tobaccos were amongst Britain's most popular brands, and their containers remain just as much in demand with today's tin collectors. John Player, the firm's founder and pioneer of the ready-packaged tobacco trade, originated the brand 'Navy Cut' — the first tobacco to be marketed with the manufacturer's name on its packaging.

The idea of adopting a registered trademark was also his and the appropriately nautical sailor and lifebelt design was evolved to go with the 'Navy' brands. The evocative and nostalgic imagery of this justifiably famous packaging trademark, complete with golden sunset and distant lighthouse, symbolized Britain's heroic ideals and patriotism. The basic image remained constant on Player's packaging for many years, although there were subtle variations in design, colouring and the use of embossing. For example, the *Navy Mixture* tins shown here, the smallest of which stand 7cm(2¾in) high, feature ships *inside* the lifebelt — a sailing ship and a steam ship, possibly a play on the words 'Navy mixture'.

Many people often wonder about the identity of the bearded sailor. Company records show that the right to use the sailor's head used in an 1880 advertisement for *Jack's Glory* tobacco was bought from W.J. Parkins of Chester in 1883. This head was used in various forms until the design was registered in 1891. After 1898 however, a new sailor appeared in the lifebelt. According to his obituary, 'a likeness first appeared in the "Army and Navy Illustrated" of 1898, whence it was borrowed for advertising' (by Player's) of Thomas Huntley Wood, who died in 1951. A friend of Wood's requested a fee of £15 from Player's, but Wood settled for 2 guineas and 'a bit of baccy for myself and the boys on board'. He also shaved off his beard to avoid recognition after continual questions and jokes.

Many of Player's tins, including the rare navy blue tin for 22 carat gold-tipped cigarettes, were made by their neighbours in Mansfield, Barringer, Wallis & Manners, and some are featured in the early photograph of Barringer's studio on page 17.

Left: A fine selection of tobacco tins described on page 67. *Pilot Flake* measures 14.5cm(5¾in) wide.

DECORATIVE PRINTED TINS

Barringer, Wallis & Manners printed these three colourful tobacco tins in about 1910. The similarities between them extend not just to their almost identical sizes — they all measure 16cm(6¼in) wide — but also to the fact that they are all crudely constructed, their lids being the only printed surfaces.

James' *Cavalry* was issued by Banks & James and shows the officer spearing a packet of cigarettes with his lance. Breathing holes for the tobacco are visible on the tin's side. On Cope Brothers' *Golden Cloud* the stars and stripes motif is once again used in connection with the Virginian tobacco contained in the tin. Gallaher's *Late and Early* displays several similarities to Hignett's *Night and Morning* shown on page 68. Both tins were printed at Barringer's and it seems more than coincidental that both should feature a sun and moon device (for 'Night and Morning' and 'Late and Early') to represent the mixture of two flakes contained in the tins.

Although these two curved containers are based on the principle of the popular hip flask, they are more probably breastpocket tins. However, since most of the usual tins for 20 cigarettes were flat and fairly slim anyway, the idea was never generally adopted by the major manufacturers. Cohen Weenan's *Afrikander Flake* measures 8cm (3⅛in) wide.

The brightly coloured pocket-sized sample tobacco tins each contained between 1 and 2 oz of tobacco. Many of them bear the same images as the larger tins of the same brand, and are impressive in that none of the detail has been lost in the scaling down.

As with the larger tins, the maritime theme maintains a prominent position. Two of the most attractive and noteworthy examples shown here are the *Victory Navy Cut* (after Nelson's Flagship) from Singleton & Cole which, like Salmon & Gluckstein, was a chain of tobacconists offering its own brands, and Fleming & Telfer's *Locker*, measuring 6cm(2⅜in) wide, which portrays the fictional pirate Davy Jones, guarding his legendary locker of treasure.

The three Hignett's *Hand Cut Virginia* tins exhibit subtle changes in design and use of wording. The central tin is the only one without a matchstrike (see page 88) on its base. It was printed at Barclay & Fry, and its depiction of the stars and stripes motif, often used on tins containing Virginia tobacco, is more detailed than on the others. Since a London tin printer was used, it is likely that these tins were issued by the London Hignett Company, rather than by Hignett's of Liverpool.

CIGARETTES AND TOBACCOS

Whether these splendid 14cm(5½in) high Player's *Country Life* tobacco tins were intended to be used together or singly, they are certainly unusual both in shape and in that, unlike most tobacco tins, the manufacturer's name is very discreetly placed. Given this characteristic, the tins may have been issued in sets for the Christmas market, but, despite the scenes of country pursuits, it is unlikely that they were produced for domestic use — unless for the bulk-buying pipe smoker. More probably they formed an impressive and decorative tobacconist's counter display, from which he could weigh up more modest measures for customers.

Player's enjoyed great success with this brand from the 1900s on. More conventionally shaped and sized packets and tins also came in red as well as blue.

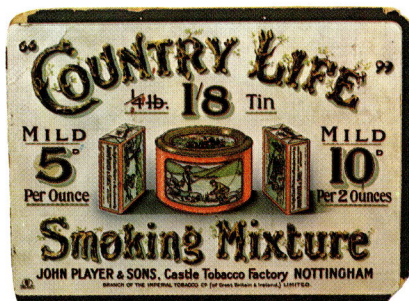

Salmon & Gluckstein made many claims for both their company and their products but, unlike those of other companies, most of theirs could be substantiated. They were indeed 'the largest tobacconists in the world' with a vast chain of retail outlets, and considered so important that the newly formed Imperial Tobacco Company felt it imperative to secure them (which they did in 1902) in an attempt to forestall the possible American Tobacco Company take-over.

Famous for their discount prices on other companies' wares sold in their shops (until controls were brought in), they applied the same policy to their own brands of cigarettes and tobaccos, amongst which the beautiful *Life Boat* brand must be one of the best known. These tins bear the ambiguous caption 'see what we save', a motto that could be applied equally well to both the lifeboat and to their products' value for money. More examples of the popular heroic nautical theme, these desirable tins feature particularly fine 'rope' lettering. The tin on the right is 16.5cm(6½in) wide.

Right: A fine selection of cigarette tins described on page 67. The *King Rufus* tin measures 8.5cm(3⅜in) wide.

CIGARETTES AND TOBACCOS

On each of the eight decorative red, blue and gold tins featured here the Muratti's lounge lizard is off in the conservatory again for a leisurely *After Lunch* cigarette. Aimed very definitely at the man about town, the style of the image-conscious gentleman on the tin certainly kept pace with potential customers, as did his background decor with current fashion. In fact, taking into account the time span of the tins of probably less than thirty years, the number of subtle changes in the design is quite considerable.

Over the years moustaches go out of fashion, suits, shirts and shoes change, a white wine bottle replaces a red one, and is in turn replaced by a decanter — the fruit on the tree even ripens, and so on. By the time the tin was redrawn in the 1930s, the style of depiction has changed completely and a coffee pot has superseded the wine, although the basic concept of the design and the red company logotype both remain the same. All the tins measure 9cm(3½in) wide; one is also shown on page 73.

This large store dispenser was issued by the Irish firm of William Clarke and it contained their *High Toast* variety of snuff. The firm had a disrupted history. Founded in Cork in 1830, they moved to Liverpool in 1870. However in 1923, when the Irish Free State became a separate customs unit, they returned to Dublin, only to become amalgamated with Wills (Ireland) in 1929. The tin stands 22cm(8½in) high.

This particularly striking 1 lb tin was issued by Hignett Brothers of Liverpool. Together with their associated company of Hignett's Tobacco Company of London, they were amongst the original thirteen firms that formed the Imperial Tobacco Company in 1901. *Cavalier* was one of the Liverpool company's more popular brands and, like most tobaccos, it was available in several sizes. A smaller example is shown on page 68. All the different sizes feature the same very splendid cavalier, pipe in hand and standing beside a vase of tobacco flowers against a tiled and friezed background.

It was a middle-priced tobacco, selling at 5s 5d per lb in 1904 (compared with Lambert & Butler's *Log Cabin* retailing at 7s at the same date). Hignett's printed inside the tins' lids a lengthy list of their product's merits, lest the smoker forget that the tobacco contained 'the maximum of fragrance and aroma' and 'the minimum of pungency and heat'. This tin measures 18cm(7¼in) across.

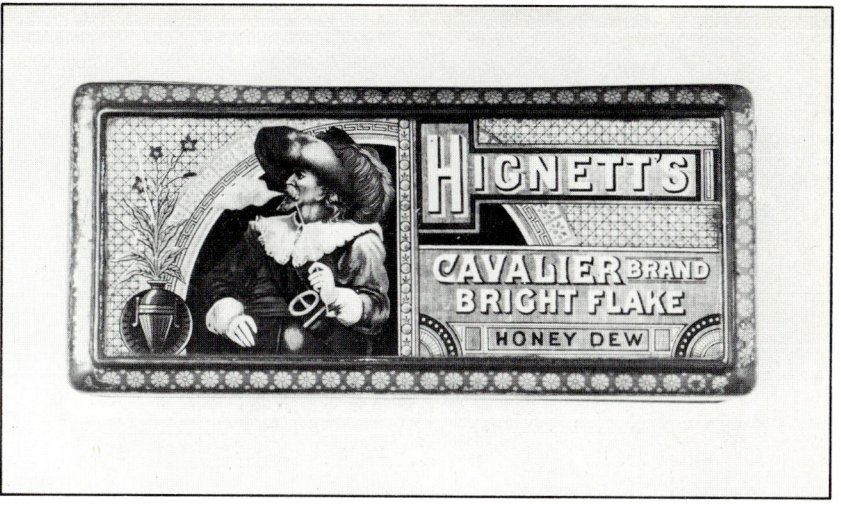

Between December 10th 1901 and October 1st 1902 Ogden's of Liverpool was owned by the American Tobacco Company. It was the first step in the plan of their president, James 'Buck' Duke, to take over the British and European tobacco trades, and was the cause of the 'Tobacco War' and the consequent formation of the Imperial Tobacco Company. It could therefore be that the poster shown here, which depicts a splendid range of Ogden's products, was produced during this time of American ownership. Certainly all the packages on it were 'manufactured for export only', and the patriotic message on the *Guinea Gold* tins, 'British made by British labor', is in American spelling.

As well as showing examples of some of their most famous brands, the poster also illustrates Ogden's different methods of marketing cigarettes and tobacco. Apart from the regular tins and packets, some of which came with mouthpieces, cigarettes were also offered 'packed under our patent cold vacuum process protected all over the world'. This process was quoted as actually improving the smoking properties of the cigarettes. Indeed so proud were they of their product that they even somewhat over-extended their use (or abuse) of literary licence by quoting Ben Jonson and claiming that theirs 'was the most sovereign and precious weed earth ever tendered to the use of man.'

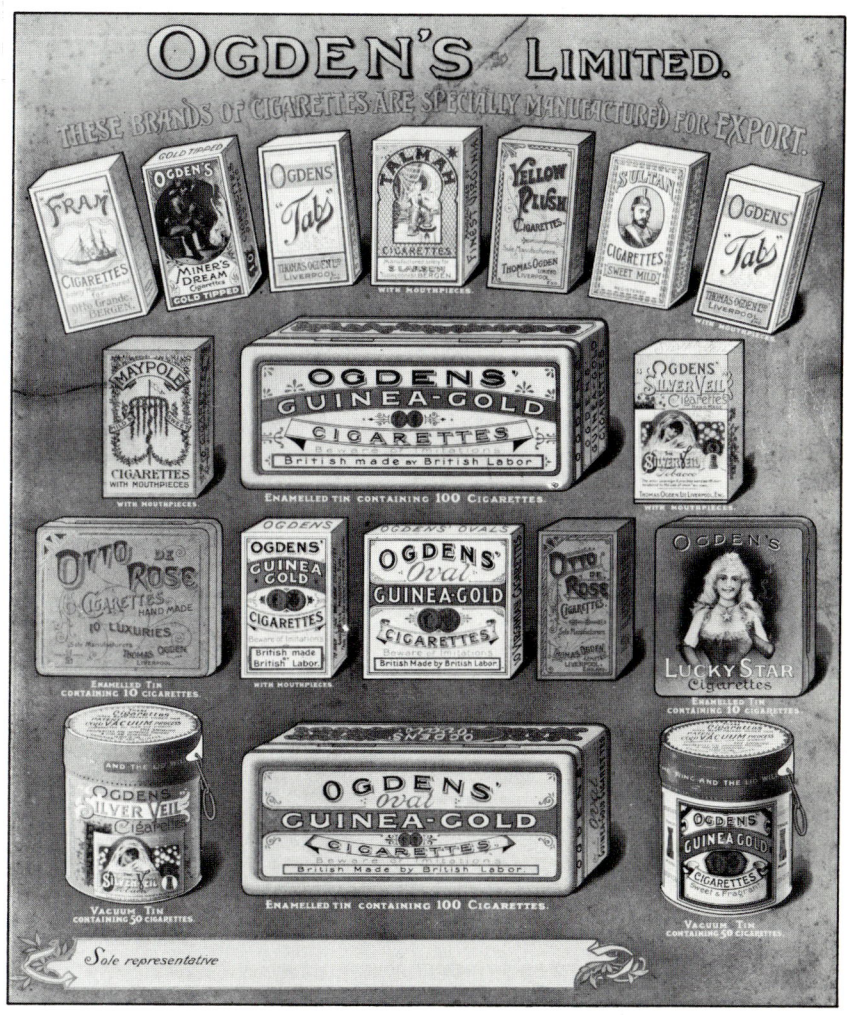

DECORATIVE PRINTED TINS

This group of finely printed tins illustrates the use of the female figure on cigarette packaging. Although at first glance the tins appear to be similar in intent, there is an interesting distinction to be drawn between them. The three earlier turn-of-the-century tins, *Gaiety Girl*, which measures 9cm (3½in) across, *Myrtle Grove* and *Woodland Belle*, all make use of the 'pin-up' as an obvious attraction to male smokers. The slightly later *Young Ladies* is ambiguous in intent; it could equally well be a continued use of the pin-up idea or it could be advocating smoking for 'young ladies'. By the mid-1920s, when *Miranda's Dream* was issued, there is no doubt. Smoking now a socially acceptable habit, the emancipated lady puffs away at one of her multi-coloured cocktail tipped 'oriental am-bar perfumed' cigarettes, dreaming of exotic places.

Both *Woodland Belle* and *Myrtle Grove*, printed at Hudson Scott, are fine examples of the early tin printer's art, particularly of hand-stippling techniques. The name *Myrtle Grove* incidentally refers not to the 'belle' on the tin, but to Sir Walter Raleigh's home, and was one of the most popular brands of the London tobacco company Taddy & Co.

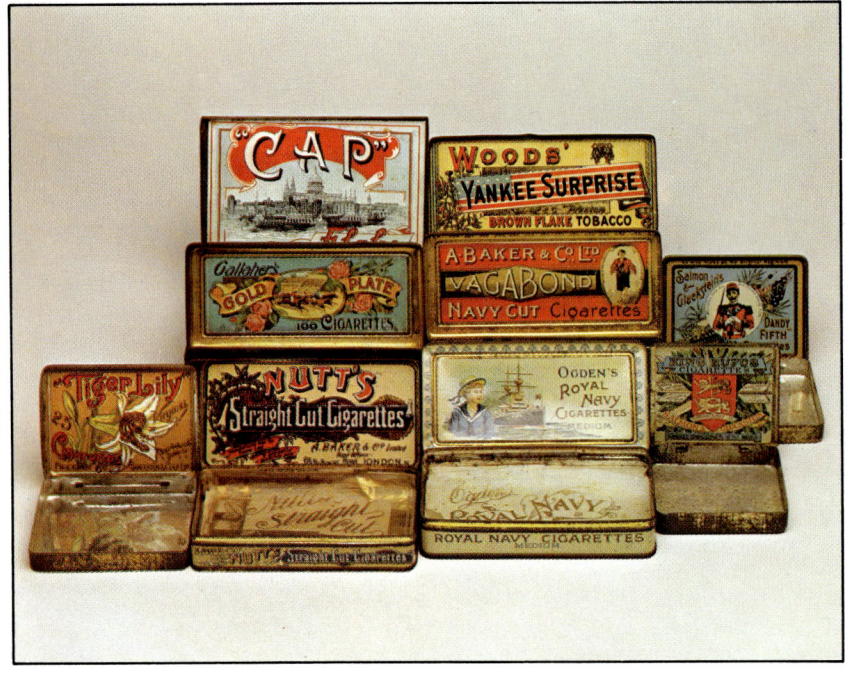

By today's standards, the lengths to which companies went in order to make the presentation of their products as attractive as possible are certainly impressive. Not content with illustrations on the outside lids, many cigarette tins also have some kind of feature inside their lids. Some bear only a simple black-on-gold illustration or utilize the space to proclaim the qualities of their contents, while others carry fine full-colour depictions such as those shown here.

In some instances, as with the *Cap* flake tin for example, the internal illustration is on a paper label rather than printed directly onto the tin. Sometimes the design is the same on both outside and inside; sometimes it is completely different, as in Salmon & Gluckstein's *King Rufus* tin, the outside of which is shown on page 73. This idea of using the inside lids for illustrations has proved a boon for collectors since, although many of the early tins are now found in poor external condition, the insides usually remain unaffected by the ravages of time. Note also the fine gold-printed paper wrappers in the tins. The *Dandy Fifth* tin ('We named them the kid-gloved Dandy Fifth') measures 10.5cm (4⅛in) wide.

COMMEMORATIVE TINS

CONTEMPORARY EVENTS of topical interest, particularly royal occasions, had been commemorated on china and glass since Queen Victoria's Coronation in 1837, almost fifty years before the first commemorative printed tin boxes. Although the date when an historic event was first committed to tin is not known, one of the very earliest must be a small Barringer & Brown (later Barringer, Wallis & Manners) mustard tin celebrating the 1875-6 Antarctic expedition, seen recently at their Rock Valley premises.

From about that date, every kind of remarkable event of national importance began to be commemorated on tin containers, patriotically printed in bold reds and bright blues, and emblazoned with portraits, flags and medals. Designs became progressively more spectacular and lavish, reaching their peak of achievement at the time of the First World War. With the one known exception of Jacob's *Coronation Coach*, designers steered well clear of novelty shapes, concentrating instead on detail and quality of the images depicted.

Few product manufacturers ignored the opportunities to capitalize on the enthusiastic patriotism of the heyday of the Empire in the late nineteenth century and, since this coincided with great progress in tin box making and printing, a vast range of products, from mustard and biscuits to tobaccos and chocolate, began to be marketed in commemorative containers. The manufacturer's name was usually kept small; on many tins it is missing altogether and the issuing company can only be identified by the tin's shape.

Subjects chosen for depiction on these tins concentrated on three main areas: royal events and personages, military exploits and heroes, and pioneering achievements, inventions and explorations. Coronations and jubilees in particular captured the public's imagination; royal weddings, the opening of Parliament or visits to International Exhibitions were also popular. In fact, almost any royal event resulted in the issue of many a new commemorative tin.

Produced in especially large numbers were the small, flat, oblong chocolate tins bearing round or oval portrait medallions, and ordered on many royal occasions by the cities and boroughs whose names appear on the tins' lids. Large cities commissioned individual designs, but smaller towns with smaller financial resources had to make do with standard tins, their names inserted in the space left blank; since many tins were given free to local children by the town authorities, expense was certainly a factor to be considered.

Vast numbers of military commemorative tins were also given away, and sent at Christmas to the troops at the Front during the South African and First World Wars. In some cases, several tin printers and box makers were involved in production, to cope with the large numbers of tins required at speed. Hudson Scott made many of the 1914 *Princess Mary* tins (see page 81), receiving, on completion of their order, a letter of gratitude on behalf of the Princess. She was apparently also 'greatly touched' by Hudson Scott's gift of a box made in silver gilt from the tools and dies used in the manufacture of 'brass' boxes for Her Royal Highness's Sailors and Soldiers Fund. Occasionally individual companies, as opposed to federations or funds, dispatched Christmas tins containing their own products to the troops. Fryer's had a beautiful Victory-V tin (see page 81) ready for issue to the general public in 1914, but in a patriotic gesture they were sent instead to the brave lads in the trenches, and many were the cards of thanks received in return.

Other military tins produced from before the turn of the century until the First World War recorded earlier campaigns in far corners of the Empire. Many combined scenes of heroism with illustrations of the generals involved, Lords Roberts and Kitchener among the most often depicted. By the Great War, Union Jack-festooned portraits had taken prominence over heroic episodes, and British commanders were joined on tin by their opposite numbers in the Allies.

Interestingly, the majority of the portrait cameos of both royalty and military leaders are very obviously based on the same original paintings or photographs, usually a standard three-quarters head and shoulders view, taken presumably from an official reference provided by the Palace or the Forces Press Office. This was in apparent contradiction to the freedom in other areas of contemporary advertising, where images of the Queen and her daughters, for example, appear quite indiscriminately promoting a wide range of products from hats to toothpowder and soap to drinking chocolate.

Thanks to their original patriotic appeal, many of these bold tins have survived to recall brave exploits and grand ceremonies, serving now as colourful reminders of the days when Great Britain ruled two-thirds of the earth's surface.

An enormous variety of commemorative tins appeared in celebration of events at home and abroad. 'Live in peace forever, part with Terra Bona never' is the motto on this *Terra Bona* tea tin (far left), issued in commemoration of the 1902 peace treaty between Great Britain and South Africa, though it is doubtful that the latter of these statements was uppermost in the minds of John Bull or his Boer War adversary, depicted on this 16.5cm(6½in) high tin.

Harvey & Davy's unusual yet appropriately shaped tobacco tin was issued for Queen Victoria's Diamond Jubilee in 1897. Measuring 7cm(2¾in) across, this decorative and detailed blue, yellow and red tin was made at the company's Hanover Works in Newcastle-upon-Tyne.

A large head-and-shoulders portrait of the distinguished Welsh-American journalist H.M. Stanley is depicted in graphic detail on this commemorative mustard tin. Somewhat surprisingly, the expedition also featured is not his celebrated discovery of Dr Livingstone, but a less well-known, yet equally dangerous foray into Southern Sudan in pursuit of Emin Pasha, the 'heir' to General Gordon.

For several years nothing had been heard of Pasha and his four-thousand strong force which had been engaged in bitter conflict with the Mahdi, whose army had slain General Gordon. So during 1886, in an attempt to establish commerce in this equatorial province, a group of British businessmen financed Stanley to find him. Some three years later Stanley, after a gruelling journey, finally found Pasha. The tin measures 23cm(9in) long.

Small flat commemorative chocolate tins were produced by confectionery companies on various royal occasions during several reigns. Coronations, jubilees and visits to particular towns were celebrated on tin, the latter issued in particularly large quantities by the local councils for the children of the area. Though Rowntree's were the most prolific of the chocolate companies to produce these oblong tins decorated with oval royal portrait 'cameos', other firms compensated for the lack of quantity by producing tins of exceptional quality. The superb 14.6cm(5¾in) long tin issued by Faulder's *'High Class' Chocolates* of Stockport (second from bottom on the left) is a particularly fine example. Commemorating the Coronation of George V and Queen Mary, the double portrait is accompanied by a depiction of the Houses of Parliament, the two illustrations entwined by an Art Nouveau border.

COMMEMORATIVE TINS

A small but representative selection of 'royal' tins, giving an idea of the variety of superb containers issued in celebration of different royal occasions. Particularly imaginative and well-produced are the Victorian tins, which often feature, as well as the obligatory portraits, detailed depictions of remarkable figures and events of the Queen's reign, such as her Coronation, on the Barringer's 22cm(8¾in) wide container, or of her territories, here in map form on Callard & Bowser's beautifully printed tin.

The pale blue Queen Victoria tin on the far right, issued by Hudson Scott, is fascinating because it bears a portrait of Prince Albert. The Queen, distraught at the death of the Prince Consort, had decreed that his Royal Warrant be removed from all products and that his image should not appear on any commercial advertising or packaging. Somehow this tin, probably produced at the time of the Diamond Jubilee in 1897, managed to slip through the censor's net.

Issued by Macfarlane Lang, the superbly embossed *Seven Edwards* tin on the extreme right features not only Edward VII, but also imaginative miniatures of his six preceding namesakes. It sold at 10s per dozen, and was filled with fancy mixed biscuits.

Four commemorative tins, produced about the turn of the century, which all contained mustard. The 4lb oblong Colman's tin on the left has on its lid a full-length portrait of Queen Victoria, as opposed to the more usual head-and-shoulders depiction; its sides are decorated with memorable historical moments during her reign, including the first passenger-carrying steam train journey between Stockton and Darlington in 1875.

Both the tins in the centre are Edwardian. The lower Keen, Robinson tin, with an embossed crown on its lid, was issued from 1901 to 1903 in celebration of Edward VII's accession. Each side carries a different portrait of the King, that visible here showing the new monarch in coronation robes set against the towers of Westminster Abbey. The tin above it was issued in 1902 for his coronation, and printed by Hudson Scott. The then Duke of Cornwall (later George V) and his extremely young-looking mother, Queen Alexandra, look out from the sides of this purple-topped Colman's tin, which stands 16.5cm(6½in) high. On the right another Keen's tin illustrates ceremonial Edwardian scenes, including the opening of Parliament to the accompaniment of a fanfare performed by the Royal Heralds on the tin's corners.

COMMEMORATIVE TINS

'Hilton's Boots are like our generals — famous for endurance' is the motto on the Baden Powell tin in this picture, and endurance was certainly the word for many generals' continuous appearance on patriotic and brightly coloured tins issued to commemorate military events and achievements over the years. Unfortunately, however, many of these tins do not carry manufacturers' marks and, where it is not possible to identify the issuing company by the tin's shape or inside label, the original contents can only be speculated on.

Of this splendid group of military tins, about half commemorate episodes and commanders of the First World War, while the others refer to the events and heroes of earlier campaigns in Africa and India. Lord Kitchener, easily recognized by his magnificent dark moustache, is depicted both on the early red bordered tin (far left) during the 1898 advance on Omdurman, as well as on all the flag-festooned tins of the First World War, when he was War Minister. He is accompanied on these tins by the other war commanders: Sir John French (Army) and Sir John Jellicoe (Navy). Do not confuse French with Lord Roberts, who appears on both the shaped Keen's mustard tin lid and the container to its left, and who died in 1914. Both men sport large white moustaches. The *Victoria Cross Episodes* tin measures 13cm(5$\frac{1}{8}$in) high.

All these tins were sent as gifts to the British forces engaged in active service during the South African War and the First World War. Of the Boer War mementos, the yellow tin is by far the most scarce, as it was issued to Scottish regiments only. Much more common is the 15.2cm(6in) long *Queen Victoria* tin, here mounted with mementos in a frame, and produced in 1900 by all three major chocolate manufacturers, Fry's, Cadbury's and Rowntree's, with subtle differences in design and printing.

Of the First World War tins, Victory-V sent out between 300 and 400 thousand of their now rare breastpocket tin to the troops. The 'brass' lacquered *Princess Mary* tin is however one of the most frequently found tins in Britain. It was literally issued in millions, and it differs from the other tins in this illustration, which, with the possible exception of the oval tin presented by the 'British Grocers' Federation' in 1914, were originally filled with a chocolate ration. In 1914 however the non-commissioned serviceman had a choice of chocolate or cigarettes and tobacco to accompany his card from the 17-year-old Princess, daughter of George V and Queen Mary. But what about the officers? For them a pencil encapsulated in a spent shellcase.

Three dramatically bold mustard tins from Keen, Robinson, gloriously overdecorated with ermine drapes and gold coats of arms. Though the exact date of their issue is not known, it is more than likely that they came out between 1900 and 1910, or perhaps just after. Clues to the dating are given by the choice of figures portrayed on the tins' sides and lids. Kaiser Wilhelm II, the German Emperor (complete with the much-ridiculed helmet), the Tsar of Russia, the Emperor of Austria and King Victor Emmanuel III of Italy appear on the sides of all the tins, but only two show Edward VII and Queen Alexandra on their lids, the third carrying a portrait of Lord Roberts.

Victor Emmanuel III only came to the throne in 1900 and by 1910 Edward VII was dead, so it seems probable that after his death the tin was in fact re-issued, but this time with Lord Roberts' picture on the lid. If this is the case, it would certainly not have been on offer for many years after 1910 since it still features the man who was to become Britain's arch enemy, the Kaiser. However the original occasion that these tins commemorate is still open for conjecture. The larger 6lb tins measure 20.8cm (8¼in) high.

One of the most interesting commemorative tins ever produced was this *Victoria Cross Episodes* oatcake tin from McVitie & Price. Its superbly embossed sides and lid feature artists' impressions of five separate acts of courageous endeavour, which won for those concerned Britain's supreme military accolade, the bronze Victoria Cross. This decoration had been instituted by Queen Victoria in 1856 and, until the supply was exhausted in 1942, all the medals were struck from the metal of a cannon captured from the Russians at Sebastopol.

This tin was produced in the 1880s, but, due to an unusual occurrence, it is thought to have had a short life. Gunner James Colliss RHA (his name was mis-spelt by McVitie & Price) was one of the VCs whose brave actions were commemorated on the tin. He had received the decoration in July 1880 at Maiwand in Afghanistan, but unfortunately forfeited it 'under royal warrant' in November 1895 on account of some misdemeanour and, given moralistic Victorian attitudes, the tin would have been swiftly withdrawn. Among the other acts of conspicuous bravery depicted on this 12.5cm (5in) high tin are 'Surgeon Jee dressing the wounded under fire at Lucknow' during the Indian Mutiny, and Sergeant Ablett in the Crimea 'in the trenches before Sebastopol throwing out a live shell'.

COLLECTIONS

DECORATIVE PRINTED TINS tend to fall naturally into several obvious categories based on the products they contained, but many of the most interesting and varied collections cut right across these divisions to reflect different individual tastes and interests. Within such a wide collecting area, the scope for building a unique collection is enormous, and the following pages are intended to give only an indication of a few of the hundreds of different themes around which a collection might be built.

While many people are drawn to tins which contained a particular commodity, such as mustard, tea or cigarettes, others prefer to concentrate on the packaging of a particular company such as Huntley & Palmers or Victory-V. The subject matter of the illustrations can also become the common factor in a collection, for example the many different depictions of naval heroism on a wide range of tins.

Novelty tins in impressive and often outrageous shapes hold a great attraction for many collectors. At the other end of the scale, advertising tins with plainer, more utilitarian shapes but with colourful, graphic depictions and bold lettering make a superb visual reference collection of typefaces and decorative motifs.

Often tins form part of much wider collections or complement those based on other specialist areas. Needle tins, for example, are ideal accessories to a horn gramophone collection, as are novelty vehicle tins to a mechanical tin toy collection. The possibilities are endless, but every individual collection of printed tins is unique.

Buses, trams, taxis, delivery vans and limousines, most complete with beautifully printed passengers and drivers: this is certainly the kind of traffic jam that every novelty tin collector often dreams of. Tins in the shapes of vehicles such as the unique group shown here are considered by enthusiasts to be amongst the most desirable of all containers — though some are in fact so rare that many collectors may not even be aware of their existence. Unfortunately the attraction of these tins was to children then, as it is to enthusiasts now, immense, and although the tins' original appeal was increased by the fact that they made superb toys, they were originally created only as transient containers and not equipped to withstand the ravages of the nursery.

The five cars, on which no manufacturer's name is visible, are from left to right: Jacob's yellow *Limousine*, a 'Toy motor car filled with choice mixed biscuits', issued in 1924; Crawford's maroon sports *Coupé*; their superb navy blue *Rolls Royce*; and, last, W. Dunmore & Sons' *Limousine*. On the front right hand is Rowntree's *Taxi Cab*. The trunk the driver carries next to his

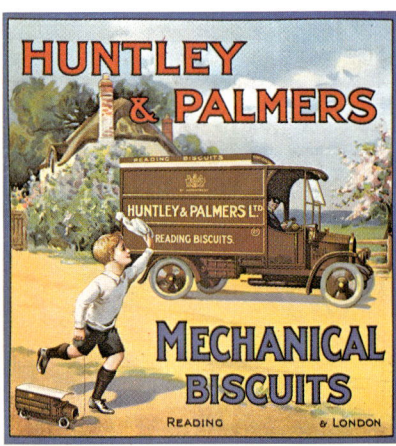

own seat is incidentally identical to the Rowntree's miniature on page 49. The roofs of all the vehicles shown here act as lids, though what was contained inside varied from biscuits to toffees to chocolates. In several instances the designers were considerate enough to incorporate the initial date of production into the design of the vehicle, on the number plate, such as Dunmore's *Limousine* of 1923, Rowntree's *Taxi Cab* of 1926, and Huntley & Palmers' 9.7cm(9¾in) long *Delivery Van* of 1923, which is also featured with its life-size counterpart in the advertisement for 'Mechanical Biscuits' above.

COLLECTIONS

No collector interested in gardening would want to be without the two tins shown here. Certainly *Garden Roller* and *Lawnmower* show the somewhat bizarre extents to which the tin designers' imaginations went in choosing even the most unlikely subjects for depiction in tin, but both these tins would be admirable additions to any collection containing packaging, ephemera or other items based around a 'gardening' theme.

The simply printed pale green and red *Garden Roller* was issued by Huntley & Palmers in 1913-14, and a flap in the main cylinder, which measures 11.6cm(4⅝in) across, acts as the tin's lid, providing access to the biscuits contained inside. In contrast, the dark red and blue *Lawnmower,* which measures 24cm(9½in) long with the handle extended, is more complicated in both construction and printing. Its detailed depiction is very realistic — from the vicious-looking blades to the finely printed pile of freshly mown grass in the hood, which forms the tin's lid. The small shield on the lid shows that the tin was made by Barringer, Wallis & Manners of Mansfield and, since there is no product manufacturer's name, it was probably one of their stock tins.

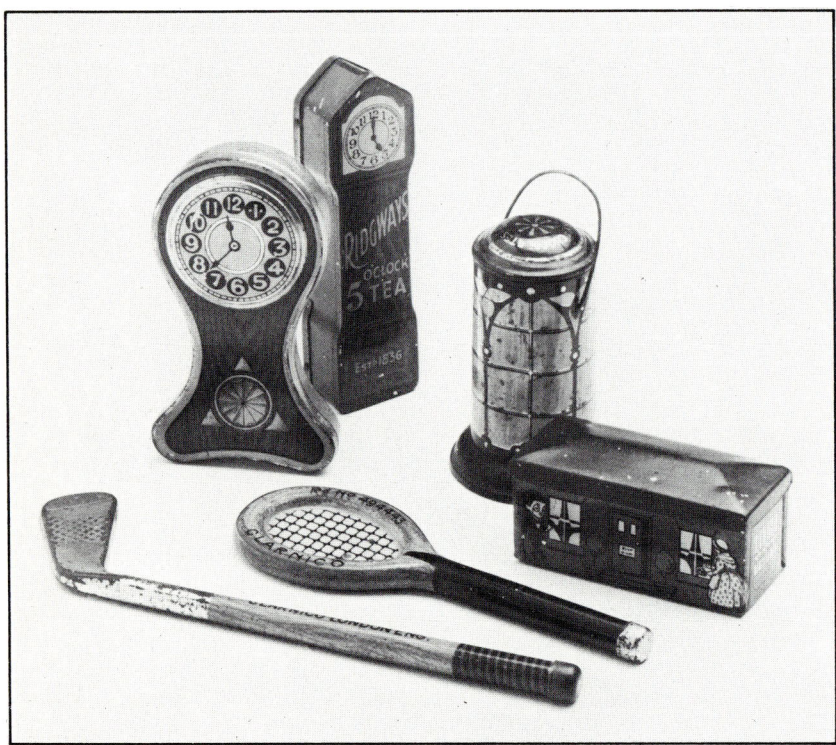

Miniature tins are particularly popular and sought after and make perfect specialist collections. Because of their minute proportions they can be easily displayed in a minimum of space, such as in one of the small advertising cabinets shown on pages 92-3 and, unlike some of the very large containers, they are easily transportable.

Many companies issued miniatures either as samples of their products or as promotional offers. Confectionery firms in particular produced them in a variety of novelty shapes, as can be seen here and in the wide selection of Rowntree's of York miniatures on page 49. Although none of those shown here are by Rowntree's, both the Clarnico *Golf Club* and *Tennis Racquet* were filled with cachous, in common with their counterparts from York. Fry's *Lantern* and their *Long Case Clock* contained sample packets of cocoa, and Sharp's *Kreemy Cottage,* which can be found in four different sizes, held a bar of Super Kreemy toffee. Predictably, Ridgeway's *Grandfather Clock,* which measures 9cm(3½in) high, contained a sample of their tea.

Tin money boxes can make interesting collections as they were produced in different shapes and sizes by quite a few companies (see also page 38), though often in the conventional red letterbox shape. This red and gold Huntley & Palmers *Penny in the Slot* is however an unusual biscuit tin money box in that it contains a mechanism housed in its lower section which releases one biscuit for each coin deposited — 'Insert a penny in the slot and pull out the tray' — thus no doubt inducing many a hungry child to involuntarily save his pennies. Basically a miniaturization of a station vending machine, the idea was used some thirty years previously by the German firm Stollwerck (see pages 92-3), though in their case the banks dispensed tiny foil-wrapped chocolate bars. *Penny in the Slot* measures 22cm(8¾in) high and sold for 1s 9d in 1939.

Although Huntley & Palmers' *Egg Stand* (top right) and their *Oval Basket* (1905) and *Work Basket* (1904-6) have no decoratively printed tin surfaces, they are nonetheless collected by Huntley & Palmers enthusiasts and by those interested in containers which were designed for specific after-use.

The 'silver' *Egg Stand* is one of the most bizarre biscuit tins ever produced — 'Whatever will they think of next?' would have been a likely response when it appeared in time for Christmas 1928. The lid and egg cups detach from the base of the tin in which the just-boiled eggs were supposed to be carried from stove to table. Its diameter is 14cm(5½in).

The real basketwork containers made ideal sewing boxes or picnic baskets when emptied of biscuits, and consequently many have become separated from their tin linings. They can quite often be found in junk shops or markets without anyone being aware of their origins, but on close inspection Huntley & Palmers' name can usually be found on one of the basket slats. Where the linings do exist, the company's name is printed on the lids in black and gold in a style very similar to that of the printed bases of the early Huntley & Palmers shaped tins (see page 23). *Oval Basket* measures 20.3cm(8in) across.

As can be seen here, not only tobacco firms issued samples for re-use as matchstrikes. Contemporary sulphur and wax-headed matches were struck on the tin's base on a series of parallel ridges, or sometimes on the company name which was impressed into the tin. This useful publicity gimmick made sure the firm's trademark received maximum exposure with frequent use of the matchstrike, although Cadbury's *Pacific Line* proves an exception since no name appears on the tin's printed surface. Quite commonly found, these small tins make easily displayed and attractive collections. Ogden's *Royal Navy Cut* measures 6cm(2⅜in) wide.

The globe was a patriotic theme popular not just with novelty tin designers and their contemporary customers, but one which still finds interest with collectors today. Shown here are a few of the variety issued by biscuit, confectionery and general provisions manufacturers. Huntley & Palmers issued the tin on the back left in 1906-7, Victory-V the one in the centre, and Crawford's that on the right. In the foreground two Rowntree's miniatures flank the World's Tea Company and Lyons' version stands on a pedestal. Although these examples are in good condition, the chances of finding a globe tin in mint condition are thin, since most were busily kept in use by families around the country following turbulent events across the world. The issue of Crawford's 1938 tin, for example, coincided with the Munich crisis of September that year. The tin's circumference is 39.2cm(15½in).

The tins which contained the old horn gramophone needles played at 75 rpm are certainly among the most popular and collectable printed containers.

The standard rectangular tins (about 3.8cm(1½in) across) contained about 200 needles, each to be used only once. As can be seen on the *All-U-Need* tin, several types of steel tips to produce varying degrees of volume were available, as well as fibre tips for the connoisseur. The famous and nostalgically familiar trademark on the *His Master's Voice* tins was often plagiarized by smaller rival firms. Francis Barraud sold his original painting of 'Nipper' his dog, together with the copyright transfer, to The Gramophone Company of London in 1899 for £100. Nipper appeared on their needle tins from about 1903, although the trademark was not officially registered until 1910, when the Gramophone Company changed its trade name to the title of the picture: 'His Master's Voice'. In the United States, 'Nipper' is used as the RCA trademark.

COLLECTIONS

DECORATIVE PRINTED TINS

It is rare that a collection of printed tins is begun on account of a conscious decision to set about acquiring a particular type of tin or a particular company's containers. Almost always it is the case of another interest or work area leading the potential collector into gradually picking up a few items which, before he realizes it, have become a 'collection' of desirable containers.

Very often the inspiration to start collecting tins is an offshoot of a business involvement, as was the case with Aldo Osti, a Paris businessman. As an executive of an international food manufacturing company whose principal commodity is biscuits, Aldo Osti is a classic example of a collector whose interest stems directly from his business. When first confronted with novelty biscuit tins, his reaction was one of amusement and curiosity. However a more serious interest developed, and since he acquired his first novelty biscuit tin he has built an impressive collection of some 200 superb items. Although he is interested in biscuit tins from all countries, 80% of his collection is British, reflecting the recognized quality of British tin printing and manufacture.

Both Adabelle Davis and Stuart and Linda Cropper work in areas of the antiques business, and it was through this that they came into contact with printed tins, and subsequently started collecting. Much of the stock of Adabelle Davis's highly successful Californian antiques shop is English, and gradually over the years on frequent buying trips to Britain she has amassed a wide-ranging collection, containing amongst many interesting advertising items some fine English mustard tins. Her rare and valuable collection of early American advertising and packaging is perfectly complemented by her British acquisitions, and the collection is superbly displayed in a 'Country Store' built right in the middle of her Los Angeles house.

While Adabelle Davis's collection includes a wide-ranging variety of tin-types, Stuart and Linda Cropper's includes a quite outstanding collection of tins based around a very specific theme. Through their dealings in early tin toys, they became interested in tins designed for re-use as toys and principally those in the shapes of vehicles. Some of their exceptional range of lorries, cars and buses contained confectionery as opposed to the more usual biscuits, but they all retain the common factor of being on wheels.

A fascination with the Boer War that began with a visit to Ladysmith in South Africa started Kenneth Griffith, actor and documentary film-maker, collecting military material, including printed tins, that related to his increasingly specialist knowledge of the Boer War. Since then his collection has extended to include items which relate not just to the South African war but also to other colonial campaigns, early explorations and discoveries and First World War items. Now probably one of the finest collections of commemorative containers in existence, it makes a superb visual reference archive of campaigns, commanders and decorations. Ironically until recently it was the tins' subject matter that was Kenneth Griffith's primary interest, rather than the knowledge that he in fact now possesses a valuable collection of very beautiful early printed tins.

It is said that Robert Opie's collection began with the wrapper of a Munchies packet. While eating the contents of this packet he realized that if someone did not start keeping and cataloguing product packaging, a whole area of daily life would disappear forever. Since then his remarkable dedication to this aim has resulted in the creation of the largest collection of packaging and advertising in existence, which relates to a vast variety of consumer products produced during the last 150 years. In 1975 the Victoria and Albert Museum, London, mounted 'The Pack Age — A Century of Wrapping It Up', an exhibition of a hundred years of packaging drawn exclusively from his collection. Included in this impressive display was of course a large number of high-quality printed tins, the varied manufacturers of these containers reflecting his wide-ranging interests, for, unlike most tin collections, Robert Opie's covers items from all of the categories discussed in this book, produced by literally hundreds of different manufacturing companies. Apart from a sound knowledge of all areas of packaging, he has studied particular companies featured in his collection, such as the Co-operative Wholesale Company. Robert Opie's interests extend beyond the containers themselves to the way they have changed our life-style — through developments in packaging, promotion and retailing.

It is hoped that Robert Opie's unique and important collection of thousands of items will soon be used as the basis of a permanent and much-needed museum or archive, which would be a valuable addition to the public collections of this country and which both it and its dedicated owner deserve.

Top left: Adabelle Davis; bottom left: Stuart and Linda Cropper; top right: Kenneth and David Griffith; bottom right: Robert Opie.

DECORATIVE PRINTED TINS

COLLECTIONS

Printed tins are only one facet of a vast variety of wall displays, shop fittings, point of sale material, packaging and ephemera all issued over the years to publicize different manufacturers' products. These different areas of advertising are all collected in their own right, but many tin collectors often choose to augment or complement their tin collections with a selection of related items.

All the objects shown here were produced within the same timespan as the tins illustrated on the previous pages — roughly from 1860 to 1940. During those years the vast majority of shop advertising was installed free of charge, though advertisers were of course assured of value for their investment since every conceivable space in the shop was available for use. Mirrors, signs and showcards covered walls, free-standing advertisements were on shelves and counters, as were ashtrays and change bowls, samples and novelties — all extolling the virtues of one product or another.

Counter-top and wall-hanging display cabinets were also provided at discount prices. 'This Cabinet is supplied at less than three-fourths cost on the understanding that it is to be used only for the display of Chocolate and Cocoa manufactured by Messrs. J.S. Fry & Sons Limited' ordered Fry's on delivery of their counter-top case. They were one of many companies to feature examples of their packaging on the 'permanent' enamel and tin signs and mirrors they used. That companies considered this a worthwhile exercise is certainly an indication of the relatively lengthy life of the package then, compared with the constantly changing containers of today.

Glass and pottery storage jars, from which small quantities could be weighed and parcelled, and large and beautifully printed — and usually unbranded — tea and coffee tins are among other items that would have graced the grocers' shelves. Novelties were also to be found there, as well as in tobacconists and toy shops. Of the five pencil sharpeners shown here, the Fry's example was a particular favourite, as was the miniature Vim kaleidoscope which stands only 5.4cm(2⅛in) high. The superb Cadbury's cricket ball is made of cardboard, a reminder that decorative packaging was not confined to tin. Even crates and boxes received the finest artistic treatment, and the illustrations for the Mazawattee tea boxes shown here were the work of the artist Louis Wain.

TIN IDENTIFICATION INDEX

A comprehensive reference list of all the tins that appear in this book, grouped under company names.

Adkin & Sons
Gold Leaf Navy Cut 73
Anvil Brand
Caramels 52–53

Baird
Coffee Essence 56
Baker (A.) & Co. Ltd.
Houris 73
Nutt's Straight Cut 76
Vagabond Navy Cut 76
Banks & James
Cavalry 70
Barringer & Brown
Mustard tin 13
Bell (J. & F.) Ltd
Three Nuns 88
Best Harper & Co.
Pioneer Tea 60
Bifurcated & Tubular Rivet Co. Ltd 56
Bisto 57
Black Beauty
Boot Polish 56
Blue Cross Tea 60
Brasso 58
Burroughs, Wellcome & Co.
Tabloid Tea 60

Cadbury Brothers Ltd
Bournville Cocoa 88
Cocoa Essence (sample) 60
(striker) 88
string tin 58
1914 Military Presentation 81
Pacific Line (striker) 88
Callard & Bowser Ltd
Butterscotch 56
First World War Generals 81, 92–93
Queen Victoria Jubilee 80
Camp
string tin 58, front cover
Carr's of Carlisle Ltd
Bus 84–85
Grace Darling Lifeboat 44
Little Romps 92–93
Tambourine 36
Carroll (P. J.) & Co. Ltd
Mick McQuaid 68, back cover
Cerebos Ltd
Bisto 57, front cover
Table Salt 66
Champion
Little Red Riding Hood 63
Churchman (W. A. & A. C.)
Silver Wreath Cigarettes 92–93
Tortoiseshell Cigarettes 73
Clarke (William) & Son Ltd
High Toast Snuff 74
Thunder Clouds 88
Clarke Nicholls & Coombs Ltd (Clarnico) 52–53
Golf Club 86
Hockey Stick 86
Horn Gramophone 92–93
Teddy Bear 8, 92–93
Tennis Racquet 86
Clarnico: see Clarke Nicholls & Coombs Ltd
Cohen, Weenan & Co.
Afrikander Colonial Flake 70
Coldstream 71
Gaiety Girl Straight Cut 76
Rattler Navy Cut 92–93
Colman (J. & J.) Ltd
Azure Blue 65
Big Game 64
'Copper' Barrel 62
'Cupboard' Tin 62, front cover
Edward VII 80
Heroes 63
Landseer 64
The Miller's Donkey 64
Naval 64
Queen Victoria 80
Seasons 64, front cover
Sèvres 64
string tin (blue) 58
string tin (red) 58
Victorian Children 64
Wedgwood 62
Columbia
needles 89

Cooper & Co.
Priory Cigarettes 73
Priory Flake 92–93
Co-operative Wholesale Society
Dainty Bits 52–53
Meadow Cream Toffee 52–53
Motor Bus 46
Syrup 57
Teapot (sample) 60
Telephone Box 41
Cope Brothers & Co. Ltd
Bond of Union 92–93
Golden Cloud 70
Crawford (William) & Sons Ltd
Berengaria 44, front cover
Bicky House 38
Bus (double deck) 84–85
Bus (single deck) 84–85
Coupe 84–85
Fairy House 38
Fairy Tree 38
Flying Scotsman 45
Globe 88
Menagerie Van 39
Regency Coach 45
Rolls Royce 84–85
Street Organ 92–93

Davies (W. T.) & Sons
Coral Flake 88
Decca
needles 89
Dominion Gramophone Records Ltd
needles 89
Dunmore (W.) & Sons
Limousine 84–85
Dunn
Chocolate Creams 56

Eagle Brand
Blood Salts 56
English (J.) & Son Ltd
Toreador Needles 89
Epps
Thunderer 92–93
Esslemans (John E.)
string tin 58
Evans & Priestman Ltd
Crusoe Toffee 52–53

Fairweather & Sons
Cut Golden Bar 71
Far-Famed Cake Co.
Greenhouse 36
Farrow (Joseph) & Co. Ltd
Who Killed Cock Robin? 18, 65
Faulder
George V Coronation 79
Field (C. & J.) Ltd
Tapers 56
Figgs & Sons
Black & Tan 71
Fleming, Telfer & Co.
Locker Navy Cut 71
Mountain Heatherdew Flake 68, back cover
Frontier Polishes Co. 56
Fry (J. S.) & Sons Ltd
Apple 48, front cover
Cocoa 88
George V Coronation 79
George V Silver Jubilee 79
Lantern 86

Long Case Clock 86
striker 8
Tourist Chocolate 92–93, front cover
Fryer & Co. (Victory-V)
Church 51
Clock Set 51
Cradle 50
Globe 88
Lodge 50
Motor Car 50
Victory Panorama 51
Warrior 51

Gallaher Ltd
Gold Plate 76
Harlequin Flake 92–93
Late and Early 70
Rich Dark Honeydew 68, 71, 88, back cover
Two Flakes 71
Gibson (R.) & Son Ltd 52–53
Gilbert (C.) & Co. Ltd
needles 89
Gospo Ltd
Catseye 66
Gramophone Company of London, The
see His Master's Voice
Gray Dunn & Co. Ltd.
Delivery Van 84–85
Our Darlings 41
string tin 58
Guardsman Records
All-U-Need Needles 89

Harvey & Davy
Diamond Jubilee Flake 78
Navy Cut 71
Henderson (S.) & Co.
Log Cabin 36
Henry
Lurline Tooth Powder 56
Hignett Brothers & Co. Ltd
Hignett's Tobacco Co. Ltd
Cavalier Brand 68, 75, back cover
Golden Butterfly 73
Golden Leaf 71
Hand Cut Virginia 71
H.C.V. 71
Night and Morning 68, back cover
Pilot Flake 68, back cover
Hill & Co.
Delivery Van 84–85
Punch and Judy Show 36
Hill (R. & J.) Ltd
Sweet and Mild 68, back cover
Hilton's Boots
Baden Powell 81
His Master's Voice
needle tins 89
Horner
Dainty Dinah Toffee 52–53
Horniman
Tea Basket (sample) 60
Hudden & Co. Ltd
Crown Navy Cut 71
Hughes Biscuit Co.
Lighthouse 46
Huntley & Palmers
Agricultural 22
Algerian 23
Blue Speciality No 1 22
Book (brown) 30–31
Book (red) 30–31

Books 30–31
Bookstand 30–31, 33
By Appointment 92–93
Cabinet 33
Canadian 22
Cannon 26
Casket (Owen Jones) 11
Casket (Sèvres) 28
catalogues 32
Cavalry 23
Chase 22
Children's Party 22
Chinese Vase 25
Christmas (Italian) 22
Christmas (Olympian) 22
Christmas Casket 25
Creel 29
Dickens 30–31
Dragon 24
Dundee Cake (sample) 92–93
Egg Stand 87
Egyptian Casket 22
Egyptian Urn 25
Elephant Bag 29
Engine 34
Farmhouse 26
Festal 22
Field Glass Case 29
Fire Brigade 23
Fruit Basket 33
Garden Roller 86
Globe 88
Good King Wenceslas 30–31, front cover
Grandfather Clock 26
Hamper 33
Handbag 33
Hunting 23
Inkstand 29
Italian Casket 22
Lakeside 22
Lantern 25
Library 30–31
Literature 30–31, front cover
Marble 28
Mexican 23
miniatures 60, front cover
Mirror 24
Moroccan Hamper 29
Motor Van 84–85
Nautical 23
Needlework 92–93
No. 1 Bronze 22
Nursery 22
Oval Basket 87
Pencil Box 29
Penny in the Slot 87
Perambulator 26
Pillar Box 25
Plates 33
Playmates 22
Queen Victoria Jubilee 80
Reticule 29
Screen 24
Sentries 27
Sentry Box 27
Sèvres: see Casket
Shell 25
Small Blue 22
Snowman Bank 92–93
Soldier 92–93
Stationery 25
Stories, 25, 30–31
Tank 34
Toby Jug 25, front cover
Tribrek Lorry 34
Trunk 29
Tugboat 34
Universal (Athletic) 23
Universal (Trunk) 33
Victorian Child and Dog 22
Wallet 29
Wallet (snakeskin) 33
Waterbottle 34
Waverley 30–33
Windmill 26
Work Basket 87

International Stores
Articulated Lorry 59

Jacob (W. & R.) & Co.
Accordion 36
Coronation Coach 40

94

INDEX

Gypsy Caravan 39
Hobby Horse, front cover
Humming Top 39
Limousine 84–85
Showboat, front cover
Sleigh 40
Trumpet 39

Keen, Robinson & Co.
Dominions 65
Edward VII 80
Edwardian 80
Field Marshal Roberts 81
Kaiser 82
Muses 64
National Game 8
Nelson 65
The Rivals 64
What Say You . . . (Shakespeare) 63
King (Frederick) & Co.
Edward's Desiccated Soup 56
Kriegsfeld (D.) & Co.
Apple Blossom 73
Sweepstake 73

Lambert & Butler Ltd
Log Cabin 68, back cover
Waverley Straight Cut 73
Lloyd (Richard) & Sons
Skipper Brand 68, back cover
London & Asiatic Tea Agency
Londasia 60
Lutona 60
Lyons (J.) & Co. Ltd
A.1. Balloon Basket 48
Egg Timer 59
Gala Night Assortment 52–53
Globe 88
Guardsman 92–93
Lantern 48
Lone Wolf 92–93
Mirror 59
Perpetual Calendar 59
Quenchie Toffee 52–53
Red Fire 48, front cover
Thermometer 59
Thunder Cloud 48
Wishing Well 8

Macdonald (D. & J.)
Kilty Brand 68, back cover
Macfarlane Lang & Co. Ltd
Anvil 38
Art Nouveau Purse 36
Basket of Strawberries 36, front cover
Bird's Nest 36, front cover
Book 46
Butterfly 42
Delivery Van 84–85
Goldilocks and the Three Bears 38
Golf Bag 37, front cover
House that Jack Built 38
Seven Edwards 80
Telephone Box 41
Violin Case 46
Water Mill 92–93
Yule Log 42
Mackintosh (John) & Sons Ltd
Beehive 54
Flying Scotsman 48, front cover
Telephone Kiosk 49
Windmill 54
McVitie & Price Ltd
Bluebird 42
Chest of Drawers 42
Lifeboatmen 92–93
V.C. (Victoria Cross) 92–93
Victoria Cross Episodes 81, 82
Victoria Cross Scenes 8
Manchester Produce Supply Co.
Spice 56
Matthews
Baking Powder 56
Maynards Ltd
Perfection 52–53
Mazawattee Tea Co.
Flowers and Butterflies 61
Galleon 61
Little Red Riding Hood (chocolate) 39
Little Red Riding Hood (tea) 61
Old Folks at Home 61, front cover (sample) 60, front cover

Period Scene 61
Meredith & Drew Ltd
Cheese Sandwich Inn 43 (sample)
Friar's Tale 92–93
Miranda 41
Miranda's Dream 76
Mitchell (Stephen) & Son
Tam O'Shanter 71
Woodland Belle 76, front cover
Morris (B.) & Sons Ltd
First World War Commemorative 81
Morris (John S.) & Son (Oils) Ltd
Monarch Lamp Oil 57
Moss Rimmington & Co.
Blue Beard 63
Famous Actresses 63
Maypole Dance 63
Naval 63
Muratti
After Lunch 73–74
Young Ladies 76
Murray
Hall Mark 71

Nectar Tea 60
Nicholls
Tom Bowling Navy Cut 68, back cover
Nuttall & Collins
St Crispin Flake 92–93, back cover

Ogden & Co. Ltd
Redbreast Flake 68, 71, back cover (small) 92–93
Rose Grove 68, front cover, back cover
Royal Navy 73, 76
Royal Navy Cut 68
St Ino Flake 71
Walnut Plug 68, back cover
Ogston & Tennants
string tin 58

Parkinson
Butter Scotch 8
Pascall
Camera 54
Peek Frean & Co. Ltd
Castle 43, front cover
Grandfather Clock 92–93
Pat-A-Cake string tin 58
The Winner 92–93
Player (John) & Sons
Country Life 'Cake' 72
Dreadnought Flake 68, back cover
Merrythought 68, back cover
Navy Cut 69, 73
(Gold Tipped) 69
Navy Mixture 69, front cover
Roundhead 68, back cover
Potter & Clarke Ltd
Slippery Elm 56

Quorn Specialities Co.
Custard Powder 56

Radiance
Devon Cream Toffee 52–53
Rex
Buttons 56
Ridgeway
Grandfather Clock 86
Riley (E. J.) Ltd
Billiards 56
R. K. Confectionery Co. Ltd
Felix Cream Toffee 54
Robinson & Sons
Terra Cotta 71
Rogers (S.) & Co.
Jackie Coogan's Creamy Toffee 92–93
Rowntree & Co. Ltd
Axe 49
Banana 49, front cover
Banjo 92–93
Bird Cage 49
Bristol Coat of Arms 88
Butterfly 92–93
Chocolate Basket 48
Chocolate Biscuit (sample) 60
Christmas Pudding 49, front cover
Coal Scuttle 49
Cricket Bat 49
Cyclist's Chocolate 92–93

Duke and Duchess of York 79
Edward VII Coronation 80
Edward VII Leeds Visit 79
Edward VII Manchester Visit 79
Eggs 92–93
Elect Cocoa 88 (sample) 60
Fire Bellows 49
Football 49
George V Bolton Visit 79
George V Coronation 79, 80
George V Manchester Visit 79
George VI 79
Globe 49, 88
Golliwog 49
Grace Darling 92–93
Grandfather Clock 49
Hammer 49
Hockey Stick 49
King Edward VII 49, front cover
Lantern 49
Mantle Clock 49
A National Beverage 92–93
Piano 49
Pie 49, front cover
Prince of Wales (Edward VIII) 79
Punch and Judy 49, front cover
Purse 49
Queen Alexandra 49, 80, front cover
Rocking Horse 49, front cover
Rugby Ball 49
Rule 49
Spinning Top 49
Straw Boater 49
Taxi Cab 84–85, front cover
Trunk 49, front cover
Violin Case 49

Salmon & Gluckstein Ltd
Birds Eye 71
Dandy Fifth 76
Gold Flake 71
King Rufus 73, 76
Life Boat Navy Cut 72, 73, 68, 92–93, back cover
Old Mahogany 68, back cover
Snake Charmer 73
Serpell (H.Q.) & Co.
Racing Car 92–93
Sharp (Edward) & Sons Ltd
Churn 48
Drum (sample) 92–93
Eaton Toffee 52–53
Kreemy Cottage 86
Kreemy Toffee 52–53
Parrot Cage (blue) 48
Parrot Cage (orange) 48 (sample) 48
Tram 84–85
Shieldhall
Heath Brown Flake 68, back cover
Sinclair (Robert) Tobacco Co. Ltd
Foursome Mixture 68, back cover
Sea Queen 70
Singleton & Cole Ltd
Perfect Treat 68, back cover
Tiger Lily 73, 76
Victory Navy Cut 71
Slade
Merry Xmas Toffy 52–53
Smith (Tom) & Co.
Cocoanut Cream Toffee 52
Stollwerck Bros
Bank (chocolate dispensing machine) 92–93
Steam Locomotive 92–93
Stotherts Ltd
string tin 58
Sylverex
Silver Toned Radio Crystal 56

Taddy & Co. Ltd
Myrtle Grove 76, front cover
Taylor Bros
Horseguards (mustard) 81
Terezol Company Ltd
Royal Metal Polish 56
Terra Bona
Boer War Armistice 78
Prince and Princess of Wales 60
Thompson (Dr) Pure Food Company:
see Potter & Clarke Ltd

Thorley (Joseph)
Egg 92–93
Thorne (H.) & Co.
string tin 58
The Tobacco Company of Rhodesia & South Africa Ltd
Matabele Mixture 68, back cover
Torond (Chas.)
Ocean Wave 71
Triumph Tea 60
Turnwright
Toffee Bucket 49
Tyler
Gold Leaf 71

United Kingdom Tea Co. 60

Victory-V: see Fryer & Co.
Vose's
Tourist Toffy 52

Walter
Palm Fireside Assortment 52–53
Palm Toffee 52–53
Warren & Co. 60
Williams
Butter Drops 56
Williams (W.) & Co.
Golden Chair 71
Willis & Roberts
Storm Cloud 68, back cover
Wills (W. D. & H. O.) Ltd
Capstan 92–93
Main Line Flake 68, front cover, back cover
matchstrike 88
Three Castles 73
Windlass 68, back cover
Wilton 57
Woods (W. H. & J.) Ltd
Yankee Surprise 76
World's Tea Co.
Globe 88
Wren (William) & Co.
string tin 58
Wright & Sons Biscuits Ltd
Cheese Sandwich Inn 43
Motorcyclist (with sidecar) 40

Yeatmans Co. Ltd
Westward Ho Toffees 52–53

Zebo
string tin 58

No company name
Aero Needles 89
Blue Stocking Perique Mixture 68
Butterfly 92–93
Chantecler Needles 89
Coal Scuttle Bank 92–93
commemorative tins, various 81
Cup Flake 76
Cupid 73
Dog and Radio Needles 89
Double String Tip 58
Dulcetto Needles 89
Elephant on Wheels 92–93
Frae Scots tae Scots South Africa 1900 81
Fullotone Needles 89
Geisha Needles, front cover
Great Auk 73, 92–93, front cover
Heart Needles 89
His Favourite Song Needles 89
Lawnmower 76
Leola Needles 89
Lord Needles 89
Marathon Needles 89
My Favourite Cachous 92–93
Nita Needles 89
Palladium Needles 89
Pegasus Needles 89
Pig and Piglets 92–93
Punch and Judy Drum 92–93
Queen Victoria and Prince Albert 80
Queen Victoria Coronation 80
Queen Victoria Jubilee 80
Stanley 78
Submarine Needles 89
Summer House (sample) 92–93
Wellington Needles 89
Winner Needles 89
Zulu Needles 89

GENERAL INDEX

Compiled by Valerie Lewis Chandler, B.A., A.L.A.A.

advertising 23, 29, 44, 54, 59, 67, 93
 brochures, catalogues etc. 9, 32, 35, 44
 code of practice 47
 inside lids 55, 75, 76
 Queen's image 77
 slogans and mottos 51, 52, 54, 59, 63, 65, 72, 75, 76, 78, 81
 within the illustration 26, 32, 40, 52, 84
 see also base of tins; company's name on tin; lids
after-'use' tins 16, 24, 29, 30, 35, 47, 55, 62, 87
 see also clock sets; money boxes; string tins; toys; vehicular tins
Albert, Prince 80
Alexandra, Queen 60, 80, 82
American Tobacco Company 75
artistic influences 16, 24, 25, 27, 28, 32, 59, 64, 78
 see also design; Jones, Owen *and* individual artists
Associated Biscuit Manufacturers 43

Baber, Henry 14, 15
Baden Powell, Lord 81
Banks & James 70
Barclay, Robert 14, *14*
Barclay & Fry 11, 14, 15, 70
Barnsley Canister Company 46
Barraud, Francis 88
Barringer & Brown *13*, 77
Barringer, Wallis & Manners 11, *13, 14, 15, 17, 20*, 31, 39, 41, 42, 46, 47, 49, 51, 69, 70, 80, 84
base of tins 23, 35, 43
basketwork (genuine) 87
Bisto 57
Boer War 77, 78, 81, 90
'book' tins 10, 30, 46
Boorne, Joseph 21
Booth, General Sir William 60
breastpocket tins 70, 81
British Grocers' Federation 81
Bryant & May 14–15
Burroughs, Wellcome & Co. 60

Cadbury Brothers Ltd 59, 81, 88
Callard & Bowser Ltd 80
cardboard packaging 26; 40, 44, 46, 47
care of tins 10
carrier cases for samples 47
Carr's of Carlisle 20, 35, 37, 43, 44
Carter, Howard 25
catches, distinctive 29, 44, 46
Cerebos Ltd 66
Champenois (Company) 64
Champion (Company) 63
children's market 35, 38, 39, 47, 48
 see also money boxes; sounding tins; toys; vehicular tins
chocolate tins 47, 48, 56, 61, 77, 78, 81, 84
Christmas market 21, 23, 28, 30, 32, 35, 47, 50, 62, 64
chromolithography 12, *19*
cigarette and tobacco tins 9, 67–76, 78
Clarke, Mary Anne 61
Clarke (William) & Son Ltd 74
Clarke, Nicholls & Coombs Ltd *8*, 47, 52, 86
Clarnico *see* Clarke, Nicholls & Coombs
clasps, distinctive 29, 44, 46
clock sets 47, 51, 86
cocoa tins 56, 86
Cohen, Weenan & Co. 70
collections 8–9, 83–93
collectors' associations 9
Colliss, James 82
Colman (J. & J.) Ltd *13, 19, 20, 20,* 62, 63, 64, 65, 80
commemorative tins 77–82
 see also royal occasions *and* individuals
company's name on tin 35, 43, 64, 72, 87
confectionery tins 9, 47–54, 86, 88, 90
Co-operative Wholesale Society 41, 42, 46, 52, 60
Cope Brothers & Co. Ltd 70
Crawford (William) & Sons Ltd 35, 38, 39, 44, 45, 84, 88

Cropper, Linda and Stuart 90
Darling, Grace Horsley 44, 63
date of production of tins 8, 21, 67, 82, 84
Davis, Adabelle 90
Densham, Mr 61
design
 changes in a pattern 56, 70, 74, 82
 copies 35
 differences 81
 see also artistic influences
direct printing 11, 12–13
 see also offset lithography
dispensers (in tins) 46, 74, 87
display of tins 25, 48, 51, 60, 72, 93
Duke, James 'Buck' 67, 75
Dunmore (W.) & Sons 84
Dunn (Company) 56

Edward VII 46, 60, 80, 82
Edward VIII 40
Edwards (Company) 56
Engelmann, Godefroye 12

fairytale themes 38, 61, 63
Far Famed Cake Company 37
Farrow (Joseph) & Co. Ltd *18*, 65
Faulder (Company) 78
Felix the Cat 47, 54
Fleming, Telfer & Co. 70
free distribution of tins 77
 see also sample tins
free gifts with sale 24, 29, 59, 75
French, Sir John 81
Fry, John Doyle *14*
Fry (J. S.) & Sons Ltd 14, 48, 49, 81, 84, 93
Fryer & Co. 47, 50–51, 77

Gallagher Ltd 75
general provisions tins 9, 52, 55–56, 88
George V 80
George, Ben 11, 13, 21
Grammar of Colouring (Field) 17
Grammar of Ornament (Jones) 15, 16
gramophone needle tins 88
Gray Dunn & Co. Ltd 35, 41
Griffith, Kenneth and David 90–91
Grocer, The q. 55
groceries *see* general provisions

handles, distinctive 42, 46, 54, 60
Harvey & Davy 78
Haythornthwaite Family 47, 50
Henderson (S.) & Co. 36
heroes theme 63, 73, 77, 82
Hignett Brothers & Co. Ltd (Liverpool) 69, 70, 75
Hignett's Tobacco Co. Ltd (London) 68, 70, 75
Hill & Co. 37
Hilton's Boots 81
hip flasks 70
His Master's Voice 88
Home and Colonial Stores 55
Horniman (Company) 60
Hudson, Paul Greville 37
Hudson Scott (Company) 11, 15, 16, 17, 20, 35, 37, 44, 47, 60, 76, 77, 80
Hughes Biscuit Co. 46
Huntley, Joseph (father and son) 21
Huntley & Palmers 9, 10, *11*, 13, 21–34, 35, 43, 54, 60, 84, 88
Huntley, Boorne & Stevens 11, 13, 14–15, 21, 23, 28, 35

Imperial Tobacco Company 67, 72, 75
International Stores 55, 59

Jacob (W. & R.) & Co. 8, 35, 37, 39, 40, 43, 77, 84
Jellicoe, Sir John 81
John Bull 78
Jones, Owen 11, 13, 15, 16, 21

Keen, Robinson & Co. 8, 63, 64, 65, 80, 81, 82

Kitchener, Lord *7*, 52, 77, 81
Lambert & Butler Ltd 75
Landseer, Sir Edwin 64
Law's Grocer's Manual 35
lids and openings
 advertising inside 55, 75, 76
 company's name 35, 43, 51, 55, 64, 87
 distinctive 24, 28, 30, 34, 37, 41, 42, 43, 45, 49, 51, 54, 59, 84
 for tobacco tins 67
 see also dispensers
lithography 11–13, *17*
 see also offset lithography
Lyons (J.) & Co. Ltd 48, 59, 88

Macfarlane Lang & Co. 35, 37, 38, 41, 42, 46, 80
Mackintosh (John) & Sons Ltd 47, 48, 49, 54
McVitie & Price Ltd *8*, 42, 82
Mann (Geo.) & Co. 15
marine and nautical subjects *see* nautical
market aimed at 67
 see also children's; Christmas
Mary, Princess 77, 81
matching sets of tins 43
matchstrike tins 70, 88
Mazawattee Tea Co. 55, 61, 66, 93
Meredith & Drew Ltd 43
Metal Box (Company) 11
military subjects 27, 34, 77, 78, 82
 see also Boer War; nautical and marine; World War I
miniature tins 47, 49, 52, 54, 60, 86, 87, 88
mirror inside lid 24, 59
money boxes 38, 49, 65, 87
Moss, Rimmington (Company) 63
moving parts on novelty tins 21, 26, 44, 46, 49, 51, 59
 see also vehicular tins; working models
Mucha, Alphonse 64
Muratti (Company) 74
music theme 37, 39, 46, 52
mustard tins *8, 9, 13,* 55, 62, 63, 65, 78, 80, 81, 82, 90

nautical and marine subjects 34, 44, 45, 46, 67, 69, *70*, 81
 see also military subjects *and* individuals
Nelson, Lord 26, 50, 65, 70
'Nipper' 88
'novelty' tins 8, 21, 24, 35, 77, 83

offset lithography 11, 14–15, 19–20, 21
Ogden & Co. Ltd 75, 88
Opie, Robert 91
Osti, Aldo 90
outsize tins 47, 50–51, 52, 62
 see also retail bins
Owen, Will 56

Pack Age – A Century of Wrapping It Up 91
packaging *see* cardboard; paper
Palmer, George 21
paper labels 21, 35, 47, 52, 55, 65, 76
paper seals 20, 64, 65
paper wrappers 20, 76
Parkins, W. J. 69
Pascall (Company) 54
Peek Frean & Co. 43
photolithography 12, 15, *18*
Player (John) & Sons *17*, 69, 72
Potter & Clarke Ltd 56
printing techniques 11–20

Ratoire (artist) 56
reproduction of early tins 7, 8
retail bins 47, 50–51, 52, 72, 74
retailing *see* shops and shopping
Ridgeway (Company) 86
Riley (E. J.) Ltd 56

R. K. Confectionery Co. Ltd 54
Roberts, Lord 77, 81, 82
Rowntree & Co. Ltd 47, 48, 49, 78, 81, 84, 86, 88
royal patronage 21
royal warrant 80, 82
royalty and royal occasions 40, 65, 77, 78, 80, 82
 see also individuals

Salmon & Gluckstein Ltd 59, 72, 76
salt tins 66
sample tins 47, 49, 60, 70, 86
Scott, Robert F. q. 62
selling *see* shops and shopping
Senefelder, Alois 12, *12*
sets of tins, matching, cumulative 43
'shaped' tins 8, 21, 23, 35
Sharp (Edward) & Sons Ltd 47, 48, 86
shops and shopping 47, 55
 see also advertising; display; market aimed at; retail bins; sample tins
simulated materials 29, 30, 32, 37
Singleton & Cole Ltd 70
'Sir Kreemy Knut' 48
size of tins 52, 67, 75, 86
 see also miniature; outsize
Smith, Dr and W. C. ('Toffee') 50
snuff dispenser 74
sounding novelty tins 37, 39, 49
Stanley, H. M. 21, 78
Stevens, Samuel Beavan 21
'stock' tins 35, 43, 65
string tins 48, 59
Sullivan, Pat 54
'Super-Kreem' toffee tins 48
sweet tins: *see* confectionery tins

Tabloid tea tins 60
Taddy & Co. Ltd 76
talcum powder tin 66
tea tins and caddies 55, 59, 60–61, 78, 86
technical processes
 constructing box 15–16, 20, *20*, 23
 printing 11–20
Terra Bona 60, 78
Thompson (Dr) Pure Food Company 56
Tin Plate Decorating Co. 11, 12–13, *13*
tobacco and cigarette tins 9, 67–76, 78
toffee tins 47, 48, 52, 54, 84
toys, tins as 35, 44, 47, 49, 52
 see also money boxes; moving parts; sounding tins; vehicular tins; working models
trademarks 21, 48, 49, 56, 60, 62, 65, 66, 69, 88
tradestamps 23, 63
transfer paper method 17–19
transfer printing 10, 11, 13–14, 19
Triumph Tea 60
Turnwright (Company) 49

United Kingdom Tea Company 60

valuation of tins 9–10
vehicular tins 34, 35, 40, 44, 45, 46, 48, 51, 59, 77
 collections 84, 90–91
Victoria, Queen 65, 77, 78, 80, 81
Victoria and Albert Museum 91
Victory-V 47, 50–51, 77, 81, 88

Wain, Louis 66, 93
Walter (Company) 52
Weierter, Louis 56
White & Pike 21
Wilhelm, Kaiser II 82
Wills (W. D. & H. O.) Ltd 74
Wood, Thomas Huntley 69
Worcester Royal Porcelain Company 25
working models 39, 40, 44, 45, 54
 see also moving parts; vehicular tins
World War I
 commemorative tins 77, 81, 91
 influence on illustration 27
 influence on production 34, 47
World's Tea Company 88
Wright's Biscuits Ltd 40, 43

Yeatmans Co. Ltd 52